The Value of Time; Behavioral
Models of Modal Choice

The Value of Time; Behavioral Models of Modal Choice

Peter L. Watson
Northwestern University

Lexington Books
D.C. Heath and Company
Lexington, Massachusetts
Toronto   London

Library of Congress Cataloging in Publication Data

Watson, Peter L
    The value of time.

    Bibliography: p.
    1. Choice of Transportation–Mathematical models.
I. Title.
HE336.C5W38   380.5   73-10010
ISBN 0-669-89326-9

Published simultaneously in Canada.

Printed in the United States of America.

International Standard Book Number: 0-669-89326-9

Library of Congress Catalog Card Number: 73-10010

To my Mother and Father

Contents

List of Figures

List of Tables

Preface

The origins of this study, like those of all research efforts, are confused and intangible. My interest in this subject was first aroused when I attempted to carry out a cost-benefit analysis for a bypass, or circumferential, highway for the city of Edinburgh, and became distressed at the state of the art of evaluating time-savings. At the same time, analysts in the British Ministry of Transport (now part of the Department of the Environment) were also becoming concerned that most of the existing values of time were from studies of commuters. Thus, the Edinburgh-Glasgow Area Modal Split Study became part of the Ministry's research program to investigate values of time in noncommuting situations. My interest in estimating values of time broadened during the life-span of the study to include mode choice analysis in general, both at a pragmatic and at a theoretical level. The results of my tortuous progress from data collection and preliminary investigations in Scotland to the complete analysis in the United States are presented below. It is hoped that they fill a small gap in our knowledge of these matters and that they may act as a stimulus for further investigation.

Clearly no study of this type can be carried out without the help of a large number of people. The Edinburgh-Glasgow Area Modal Split Study is no exception, and I cannot hope to enumerate and thank all the people who helped me. However, some provided assistance far above what I could have expected.

My greatest debt of gratitude must go to the late Neil Mansfield, who gave an immense amount of time and effort to monitoring the study for the Ministry of Transport, and who was constantly available to advise and criticize. Without his help this study would never have reached fruition.

To his colleagues at the old Ministry of Transport I am also indebted, both for the grant that made data collection and analysis possible, and for numerous and invaluable consultations. I must in particular thank Tony Harrison, David Quarmby, and Steve Taylor.

If the Ministry of Transport was willing to support me, it was at least in part owing to the support I received from the Department of Economics of the University of Edinburgh. To my supervisors, Professor I.G. Stewart and W.D.C. Wright, and to Professor J.N. Wolfe I am particularly grateful; to all the other members of staff and students who listened patiently to my dreams, many thanks.

During the data collection and processing stages, numerous agencies were involved. I am grateful to all for their help. Special thanks are due to Ian Lockie of British Rail and John Nimmo of the Social Science Research Centre at Edinburgh University.

Since my move to Northwestern I have had the benefit of discussing my ideas with numerous people. In particular, thanks are due to Professor Peter Stopher, who has patiently explained the obvious so many times, to Professor Richard Westin for his help with the mathematics, and Dr. Thomas E. Lisco, whose argumentative nature has led me to rethink and clarify numerous issues.

Clearly, this study was not organized singlehanded. Mrs. Jennifer Barnes was my assistant for two years, and I am indebted to her for her assistance. I am also grateful to the Staff of the Transportation Center at Northwestern University for their help in the latter stages.

From a monetary point of view, I must thank the old Ministry of Transport for entrusting me with the study and for financing the data collection and analysis. Some of the later analysis was carried out under the Transportation Center's research and training grant from the Urban Mass Transportation Administration. The results, however, should not be interpreted as the official position of either agency. All errors are, of course, the responsibility of the author.

Finally, I must thank my wife, Evelyn, who has shared with me all the elations and disappointments of this project.

1 Introduction

This study concerns the analysis of the time factor in transport investment, that is, the evaluation of the time that people spend traveling. Two recent trends explain the value of such a study.

First, methods of transport are becoming more and more sophisticated, with the result that people are traveling more each year. This has been a feature of society since the early Middle Ages, but the advent of the railways and the internal-combustion engine has accelerated the process. Moreover, since the Second World War, the continuing improvement of transport services and rising incomes for the majority have led to an even faster increase in the amount of travel undertaken. Increasing income and the popularity of the automobile have caused rail traffic to fall or remain constant, while road traffic increases rapidly. Table 1-1 shows the relative movements in the volume of traffic by different forms of transport over the decade beginning in 1956.

This rapid increase in the volume of traffic using the road network has led to increased investment in the road system both by central and local government. Tables 1-2 and 1-3 show the rapidity with which this investment has grown.

Table 1-1
Transport Statistics, Estimated Passenger Mileage in Great Britain

| | | | Thousands of miles | | |
| | | | Road | | |
Year	Air	Rail	Public Service	Private Transport	Total
1956	0.3	24.5	48.6	59.5	132.9
1957	0.3	25.9	45.9	59.9	132.0
1958	0.3	25.5	43.4	72.9	142.1
1959	0.4	25.5	44.1	82.1	152.1
1960	0.5	24.8	43.9	88.9	158.1
1961	0.6	24.1	43.1	97.1	165.5
1962	0.7	22.8	42.4	103.7	169.6
1963	0.8	22.4	41.5	110.5	175.2
1964	0.9	23.0	40.3	125.5	189.7
1965	1.0	21.8	37.6	132.8	195.2
1966	1.1	21.5	36.3	145.1	204.0

Source: *Annual Abstract of Statistics*, H.M.S.O., 1967 Table 233, p. 194.

Table 1-2
Expenditure on Highways—Major Improvement and New Construction from Central Government Funds

Year	Trunk*	Classified	Unclassified	Total
1955	n.a.	n.a.	n.a.	n.a.
1956	8.2	6.9	–	15.2
1957	11.9	10.5	–	22.4
1958	32.7	15.3	–	47.9
1959	42.7	17.4	–	60.1
1960	38.9	22.1	0.5	61.5
1961	47.3	24.9	0.5	72.7
1962	66.7	27.1	0.3	94.1
1963	76.1	35.2	0.7	112.0
1964	102.3	40.6	0.7	143.6
1965	101.5	40.8	0.5	142.8

*Including motorways
Sources: *Annual Abstract of Statistics*, H.M.S.O., 1967, table 234, p. 195; *Highway Statistics*, H.M.S.O., 1966, tables 28, 29(a), and 29(b), pp. 50-52.

This increase in the amount of investment undertaken by government and other non-profit-making organizations was also to be noted in other fields, for example, flood control, and the increasing governmental responsibility for investment led to some concern over the establishment of priorities. Since, by their very nature, the projects undertaken were not subject to pricing in the market, difficulties arose over the value of the products resulting from these investments. In an attempt to solve this problem of investment appraisal, the technique of cost-benefit analysis was developed.

Cost-benefit analysis can be defined as an aid to decision-making in the public sector; it attempts to make economic assessments of projects for which the market mechanism may not provide an optimum allocation of resources, because of the difficulties of deciding who the beneficiaries are and to what extent they benefit. It is not possible for private enterprise to undertake such projects, since the "product" cannot be sold in the normal way. The cost-benefit analyst attempts to enumerate and evaluate benefits, to whomever they may accrue, and costs, by whomever they may be payable. An agreed criterion is then used to make an economic judgment on individual projects or to rank projects according to their economic worth. The main aim of this type of analysis is to reduce the random element of decision-making by providing the decision-maker with systematically presented information.

Table 1-3

Expenditure on Highways—Major Improvement and New Construction from Local Authority Funds

Year	Trunk*	Classified	Unclassified	Total
1955	n.a.	n.a.	n.a.	n.a.
1956	–	5.5	2.8	8.3
1957	–	6.5	2.5	9.1
1958	–	9.5	3.0	12.5
1959	–	14.8	4.2	19.0
1960	–	14.1	6.1	20.2
1961	–	19.2	6.9	26.1
1962	–	17.2	7.1	24.4
1963	–	23.3	9.2	32.5
1964	–	28.4	10.6	39.0
1965	–	30.0	8.8	38.8

*Including motorways

Sources: *Annual Abstract of Statistics*, H.M.S.O., 1967, table 235, p. 195; *Highway Statistics*, H.M.S.O., 1966, tables 28, 29(a) and 29(b), pp. 50-52.

Much pioneering work in this field was carried out in the United States in water resource development, when the problem of assigning and evaluating the costs and benefits from government-sponsored irrigation projects arose. It had become clear that only an official agency could undertake the construction of such projects, whereas the benefits, such as flood-protection, accrued to many different people. The technique of cost-benefit analysis was developed in response to the difficulties of evaluating these projects.

It was only a matter of time before this type of analysis was used to assess projects in the field of transport. It could be used in particular to justify individual projects, or more generally to choose between competing projects.

One problem that has proved intransigeant to cost-benefit practitioners in the transport field is the question of the value of travel time or, to be more precise, the value of travel time saved. Several studies have shown that in cost-benefit analyses of transport projects, such as roads and bridges, time-savings can account for, in the lowest estimate, 25 percent of total benefits[1] and, in the highest, more than all other benefits.[2] The benefits from a project are highly sensitive to the value placed upon the time-savings in order to convert them to a monetary, and hence comparable, basis. This study is concerned with the value that is placed upon travel time-savings.

The Aims of This Study

This study attempts to examine the various methods of evaluating time which have been suggested in the last fifty years and to assess their theoretical and practical worth. The more important methods are then compared, using data from a survey of interurban transport in the Central Lowlands of Scotland.

The procedure adopted is to consider the historical development of values of time, culminating with the new generation of behavioral models. As a result of this examination, the value of time is defined in terms of traveler's choice of travel mode. Since most behavioral work has been undertaken for commuting trips, the primary aim of this study is to extend the range of knowledge to include social and recreational trips of a greater length than the typical commuting trips. Thus, the study examines trips made between Edinburgh and Glasgow and their respective catchment areas, in other words, medium-range intercity trips, and the analysis is restricted to trips whose purpose is neither business travel nor the journey to work, i.e., mainly social and recreational travel.

A subsidiary aim is to examine the methods of analysis that can be used to estimate models of mode choice, and the statistical properties of the main methods are considered before a choice is made.

Finally, the study attempts to examine the effects of income on mode-choice models.

Plan for the Book

The book follows the pattern outlined below:

The remainder of Chapter 1 will consider the historical development of values of time.

Chapter 2 sets out the behavioral hypothesis that forms the basis of the model and states the assumptions necessary to the model. It examines the form of the relationship to be modeled and demonstrates the derivation of the value of time.

Chapter 3 sets the behavioral model of modal choice in the framework of a generalized choice model, based on utility theory. It demonstrates the way in which explanatory variables enter into a choice decision. Finally, it examines the nature of the value of time derived from the behavioral model.

Chapter 4 considers the variables to be included in the model, with particular respect to the forms and combinations of variables to be used.

Chapter 5 describes the methods used and the problems encountered in collecting the data.

Chapter 6 examines the data which has been collected.

Chapter 7 examines the statistical properties of the methods of analysis available for the estimation of mode-choice models. The methods are regression analysis, discriminant analysis, logit analysis, and probit analysis.

Chapter 8 presents and discusses the results of the analysis.

Chapter 9 summarizes the conclusions drawn from the study.

Previous Attempts to Evaluate Time

While the theoretical aspects of techniques like cost-benefit analysis have been relatively well covered by economists, the practical details have been left principally to highway planners and traffic engineers, with the result that the student in this field is obliged to extend his scope to encompass the literature appropriate to these and related fields.

To reduce the findings of such a literary search to the more manageable proportions appropriate to this study, a chronological consideration of the methods used to evaluate time has been eschewed as time-consuming and tedious. Moreover, such a literature search has recently been carried out by Haney (27). Thus, a more direct procedure has been adopted. The methods have been classified under a number of subheadings corresponding to the different approaches to the subject. Each approach will be considered in general terms, and the more interesting and representative exponents will be examined.

"Assumed" Evaluations

Into this section fall all the evaluations not based on empirical research but presented as being "a reasonable view" or "representative of current opinion." In fairness to their exponents it must be said that in the majority of cases, it is made quite clear that the suggested valuation is an "assumed" valuation. In the absence of better or more reliable information, cost-benefit analysts, realizing the importance of time-savings, have felt obliged to invent or "assume" a value for the time saved; moreover, in the hope that their valuation will be widely accepted, they have, on the whole, avoided extreme valuations. Nevertheless, the variation between the different estimates has been considerable. American estimates range from the U.S. Bureau of Public Roads' $3.00 per vehicle hour (1925), to the American Association of State Highway Officials' $1.55 per vehicle hour (1960). These valuations are from semiofficial bodies, and the estimates of individual researchers vary even more.

The majority of British analysts have avoided simply assuming values for time saved. The approach of the M1 motorway study in providing multiple estimates of benefits given various evaluations of time-savings is a clear improvement over providing a single estimate, since, in the absence of an accepted value, such a sensitivity analysis is useful to the planner. Nevertheless, British analysts have assumed time values, and 2s. 6d. per hour was a common value in the late 1940s and 1950s. The Road Research Laboratory, in their publication *Research on Road Traffic* (18), also recommended an assumed valuation, calculated at 75 percent of the wage rate.

This type of valuation is reminiscent of some of the earliest attempts at valuation in which time was valued at the wage rate, on the grounds that if people sold their time at that rate, they would be equally willing to buy time at the same rate. This type of reasoning has, however, fallen out of favor in more recent times, and valuers have usually assumed that a "reasonable" valuation should be less than the wage rate. Valuation at the wage rate has been reserved for working time saved, a procedure almost universally accepted,[3] while the controversy over leisure time still goes on.

The idea that time should be valued at more than the wage rate has never received popular currency, despite the fact that the overtime rate is the rate that must be paid to induce people to give up their leisure time. This view is supported by Bellis (7). It is, however, true that, as a result of collective bargaining in the wages field, the additional payments received for overtime working may be considered as an economic rent generated by the power of the organized labor. This argument is supported by the findings of empirical studies which do not support the view that time spent in leisure pursuits is, in fact, valued more highly than working time. Moreover, as pointed out by Glassborough (25), payment is made for both time and effort, and, he claims, "little value is placed on time and high price on effort."

When the U.S. Bureau of Public Roads recommended in 1925 that time be valued at $3.00 per vehicle hour, the absence of data on the topic justified this procedure. Since then, much work has been done in the field, and although no firm conclusions have yet been reached, it is arguable that it is no longer justifiable to simply "assume" a value of time. If such a judgment must be made, it should be made on the basis of relevant empirical work, which is now available to cover many situations.

Valuations Based on Operating Costs and/or Toll Fees

In the early 1940s, H. Tucker (79) reported two methods used to measure the value of time. The first argued that even if a car was held up in a stream of traffic, the fixed operating costs still applied; if the fixed cost per mile was multiplied by the average speed, a cost per hour would be determined. This appears to be the first attempt to introduce a valuation of time based on operating costs, albeit a rather crude one. Tucker mentions it briefly and then moves on to consideration of a method based on route choices. It was noticed that drivers are sometimes willing to take a route involving additional mileage in order to save time. The value for time was calculated by calculating the cost of the extra distance and dividing it by the time saved:

Value of time = $\dfrac{O \, \Delta \, m}{t}$

O = operating cost per mile
m = mileage
t = time

This approach was duplicated by Hennes (32) in 1956. A similar method was developed by West (86) in 1946 after he noticed that traffic was generated by the opening of a toll bridge. His argument was that people were buying the time saved at the price of the toll fee. The value of time was therefore calculable as follows:

Value of time = $\dfrac{f}{\Delta t}$

f = toll fee
t = time

The operating-cost method was modified in 1958 by Vaswani (81), who split the costs into highway costs and user costs. (Unfortunately, from available information, it is not possible to deduce exactly how these are defined.) The value of time is then defined as the marginal rate of change of costs per unit of time and is expressed in terms of a differential

$V = \dfrac{\partial \, (H + U)}{\partial t}$

H = highway costs
U = user costs

In 1959, Dawson (14) working with the Road Research Laboratory, combined the operating-cost method and the toll-fee method and developed a method in which operating costs and toll fees jointly determined the value of time:

$V = \dfrac{f + O \Delta m}{\Delta t}$

(symbols as above)

This attempt to include various elements of total cost was followed up by Cherniak (10) in his analysis of travel impedances, which were represented in the following way:

$TC = O \Delta m + V \Delta t + \Delta f + e$

TC = total costs
e = residual impedances

The model is solved by the least-squares method.

In a similar treatment by St. Clair and Lieder (63), an allowance for comfort and convenience was included for the first time. These variables were measured by the number and amount of changes in speed resulting from the route under consideration:

$$\Delta TC = O \Delta m + \Delta a + \Delta f + V \Delta t + W \Delta y$$

a = accident costs
y = speed changes
W = value of discomfort and inconvenience in $ per m.p.h. of speed change

Finally, this model was extended by Claffey (11), who carried out test runs on toll and free roads and collected data on highway use, operating costs, toll costs, time, speed changes, and accident costs. The model was

$$\log r = -e \left(O \Delta m + \Delta a + \Delta f + V \Delta t + W \Delta y \right)$$

$$r = \frac{P}{100 - P}$$

P = percentage choosing toll road

This can be used to evaluate the values for time and discomfort and inconvenience by multiple-regression methods.

The methods based on operating costs alone are subject to two main objections. First, they assume that the driver is aware of his operating costs and of the mileage involved in different routes. It is by no means certain, however, that motorists do in fact know what their operating costs are. While they take account of the costs of gasoline, they rarely include wear and tear, maintenance, taxes, insurance, and depreciation (to the extent that they are use-dependent) in their calculations. It is also likely that, especially for short journeys, motorists do not accurately perceive small differences in mileages.

Second, this method of evaluating time depends heavily on the assumption that motorists base their decisions exclusively on cost, time, and mileage difference. While these variables appear to be the most important influences on the motorists' decision-making process, other factors such as safety, convenience, or comfort should be taken into account.

Later attempts to build these factors into the models are a definite improvement on their less sophisticated predecessors. They are still based, however, on relatively simple models, and only in recent years have better models been developed.

Disaggregate Behavioral Models

The next set of models have been named disaggregate behavioral models: disaggregate, because they deal with individuals; behavioral, because they are concerned with the behavior of the individual, in contrast to the models of the previous section which, although tested with the times and costs of individuals, investigated the relationships between nonbehavioral variables.

The behavioral model attempts to explain the choice behavior of the individual and has usually considered choice between travel modes or between travel routes. The model seeks to explain the probability that a traveler will choose a given mode in terms of a number of system and user characteristics. The system characteristics involve such variables as the times and costs of the different modes, and the user characteristics comprise the socioeconomic characteristics of the traveler and his family.

Such models have been tested using a number of statistical techniques, most notably regression analysis, discriminant analysis, probit analysis, and logit analysis. Having estimated the coefficients of the model, it is possible to derive a value of time from a ratio of the time and cost coefficients, which indicates the change in one just sufficient to compensate for a unit change in the other, leaving the choice probability unchanged. Such studies have been carried out by Stopher (66), (68), and Quarmby (58) in the United Kingdom, and by Lisco (42), Thomas (75), and Warner (82) in the United States. It should be noted that the aim at this point is to present the study background, location, and aims; the theoretical aspects of each study will be taken up in Chapter 3.

Before proceeding to a discussion of the models of the authors noted above, it is appropriate to pause for a moment to consider the work of Beesley (6). His work does not, strictly speaking, fall into the category of disaggregate, behavioral, stochastic choice models, since he does not, in fact, estimate a model. Nevertheless, his statement of the concept of the time-cost trade-off behavior of individual travelers forms the basis of all the subsequent models.

Working with the specific aim of estimating a value of time for a set of London commuters, he identified those travelers who could be classified as time-cost "traders," a "trader" being defined as a traveler for whom

$$a > x \text{ and } b < y \qquad
\begin{aligned}
a &= \text{preferred mode time} \\
b &= \text{preferred mode cost} \\
x &= \text{rejected mode time} \\
y &= \text{rejected mode cost}
\end{aligned}$$

The journey time and cost differences are then plotted (on what has come to be known as a "Beesleygraph"), from which maxima and minima for the value of time can be derived, depending on whether or not the traveler is gaining time or money. The value of time for the group is then obtained as the value which best divides up the maximum and minimum values (i.e., which minimizes the number of misclassifications). Thus, although no model is estimated statistically, the trade-off method is clearly established, leading the way for other analysts to develop the process further.

Stopher's study began with a small-scale survey of administrative and academic staff at University College, London. It was intended to indicate the variables that should be included in the model and to make an initial calibration. The location was selected because it was close to central London and thus well served by public transport and convenient to the radial routes into London. It was also useful that the subjects displayed higher-than-average car ownership. The most important choice-influencing variables were found to be speed, convenience, cost, and comfort. Since convenience and comfort could not be quantified, cost and time were selected as variables to be included in the model.

The model itself was intended to relate the choice between car and public transport to the times and costs of the journey by each mode. Time and cost differences were selected as the most appropriate forms of the variables and the model took the form:

$$P = \alpha\,(\,C_2 - C_1\,) + \beta\,(\,T_2 - T_1\,) + \delta$$

where P is the probability of choosing the car, C is the cost, T is the time, 1 indicates a car variable, and 2 indicates a public transport variable. δ is a constant term that expresses the probability of using a car when the times and costs are equal and will take the value 0.5 if the probability depends on time and cost alone. Any departure from $\delta = 0.5$ indicates a bias for or against the car on grounds other than time and cost, possibly comfort and convenience. Reasonable correlation coefficients were obtained from tests of the model, and resubstitution of the data reproduced the actual mode choices with a high degree of accuracy.

The model was retested using a different set of data from a second work-trip survey at County Hall, the central office of the Greater London Council. The subjects in this case were close to the average incomes and car ownerships for the city, and it was discovered that the model did not perform as well. A breakdown into income groups, however, revealed that, as was expected, the probability of using the car varied with income. The model was therefore generalized to take this factor into account, and the coefficients were expressed as functions of income.

Stopher's form of the model suffers from one serious fault. Being a linear relationship, it is possible that large positive or negative values of cost and time difference would yield probabilities in excess of one or less than zero. In an attempt to yield a more behavioral and more mathematically satisfactory model than the original linear form, Stopher transforms the model into a simple logistic form, where P is equal to $\dfrac{e^Y}{1 + e^Y}$ and Y is a linear function of the various variables.

Quarmby's work was based on a survey carried out in Leeds, England, of employees in a number of work places in the central business district. These work places were chosen to provide a reasonable spread of walking distances within the central business district. Data was collected from the subjects on the journey made on a given day, and detailed information on times and costs was collected where possible. Time and cost were the only mode characteristics involved in the analysis as Quarmby, like Stopher, felt that variables of the comfort and convenience type were unquantifiable. Quarmby also included a number of variables reflecting the socioeconomic characteristics of the subject and his household. Quarmby's analysis proceeds by way of developing the Beesley analysis[4] into a discriminant analysis form, and he derives the basic model form of $Z = \lambda_1 X_1 + \lambda_2 X_2 + \ldots \ldots + \lambda_n X_n$ where Z is the relative disutility of the public transport mode, the λ_i are weighting coefficients and X_i are relative measures of factors such as time and cost. Quarmby then saw the task of his analysis as answering a number of questions:

1. Which type of formulation (ratios or differences of time and cost) provides the best explanation of the observed mode choices?

2. What can be said about how people cost the running of their cars?

3. What is the relative importance of the various factors in affecting choice of mode?

4. How well can the discriminant function explain the choice of mode in terms of the best set of factors discovered?

5. What other relationships are there between factors, such as might affect the validity of predictions?

6. What can be said about how people value time?

7. Are there significant differences in the result obtained between car-bus and car-train choices.

Questions 1, 3, 4, and 6 are the questions of interest to this study. Quarmby found that the difference formulation of times and costs was preferable to the others. From the point of view of this study, it was interesting to note that he found overall travel time difference, excess travel time difference, cost difference, and income to be the most important variables in his analysis. It should be noted at this point that Quarmby's attempt to derive the model, in terms of the disutility of travel is one of the first attempts to provide a theoretical justification for the model. This topic will be taken up in greater depth in Chapter 3.

Lisco's study is described in his introduction as an attempt to put a value on the time spent by commuters during their daily travel to and from work. More specifically, it is a study of the marginal value of commuter travel time, a study of how much it is worth to commuters to save given amounts of time on the commuting trip. The commuting trip in question is one from the Chicago suburbs of Skokie and Morton Grove to the central business district, and the gross sample is made up of 2000 households from those towns. The information derived from this survey was used for all variables except the times and costs themselves, which were derived on an engineering basis. In other words, independent data was collected in order to make an estimate of the travel time by each mode for each observation. Using these minutely derived times and costs for alternative trip modes, the study inferred from actual choices made by commuters how they value the various factors entering into their choice decision.

Lisco developed this model by considering the response, in terms of the number of drivers choosing a given mode, to trading-off given amounts of time or cost against a fixed background of other variables. This response is derived as a cumulative normal distribution. The statistical tool normally used to fit cumulative normal curves is probit analysis and has largely been used for biological assay, but usually at a one-variable level. Lisco fits his model using the form of probit analysis that allows for multiple independent variables.

The study carried out by Thomas for the Stanford Research Institute differs from the three studies considered so far in that it was a route-choice study. Eight locations in the United States were selected which fulfilled the following three criteria:

1. A large plant, government installation, or closely spaced complex of small facilities located near an exit from a toll road. More than 2000 people should be employed there.

2. A free road should roughly parallel the toll road in at least one direction from the plant exit or provide substitution service for at least one exit, preferably several.

3. At least one medium-size population center should be near the parallel road, one or more exits away from the plant. This provides a high likelihood that the plant will enroll a number of employees from that area and that the toll road will represent a time-saving for the commuter.

Detailed data was then collected on a large number of route variables from both the toll and free roads and also on the subject's socioeconomic characteristics and attitude towards travel and the roads in question. The route-choice model was required to estimate the motorist's choice between the toll road and the free road, and its mathematical formulation was based on the logistic function, which is a probability function constrained between zero and one (cf. Chapter 7). When the motorist chose the toll road, the observed probability was equated to 0, and when he chose the free road, the observed probability was equated to 1. When the estimated probability was equal to or greater than 0.5, the motorist was assigned to the free road, and when it was less than 0.5, he was assigned to the toll road. The estimated probability was also interpreted as a percentage split that would be predicted for a group of motorists with the same characteristics. The most important explanatory variables were found to be the toll, the difference in travel time, and the income category of the motorist.

Warner's study set out to consider the intraurban modal choices made by individuals and proceeds to examine the influences on consumer choice of three economic variables (trip time, trip cost, and income) in conjunction with a number of subsidiary variables. The study is concerned with average or typical choice behavior, i.e., with explaining stochastic or probabilistic choice. In other words, the intention is to estimate the probability that an individual will choose a given travel mode. The problem reduces to two parts. Given a choice between mode A and mode B, what is the probability of choosing mode A rather than mode B? And how do the relative time, cost, or traveler's income affect this probability?

The data was collected in a survey carried out by the Cook County Highway Authority in cooperation with the Chicago Area Transportation Study and was obtained by interviewing adult members of households within the sample. Information was obtained concerning both basic household characteristics and the specific intracity trips made by those interviewed. A large part of Warner's thesis is concerned with the estimation methods for binary choice, where a binary choice is defined for a certain population by identifying each member with one of two mutually exclusive responses. Sample observations are then drawn, each of which provides data on a number of explanatory variables and knowledge of the choice made.

The problem is to estimate the relationship between the explanatory variables and the choice made. Warner utilizes two basic approaches: the discriminant approach and the regression approach. The discriminant approach involves extending normal discriminant analysis by assigning probabilities to the discriminant function. This method has the advantage of allowing the discriminant function to be expanded to a continuous function of the explanatory variables. In other words, the probabilities themselves can be taken as a function of the explanatory variables. The regression approach involves interpreting the choice problem in a manner analogous to regression problems, where the values of the independent variables are taken as fixed and an estimate is made of the way in which some continuous variant depends on the values of the independent variables. Here the problem is to estimate the probabilities for fixed values of the explanatory variables. Problems associated with probit analysis lead to Warner's using a logistic function to estimate the relationship. The choice of analysis method is dealt with later in this study (Chapter 7).

A number of other studies utilized similar methods to estimate modal-choice models. Most notable of these are Lave (41) and McGillivray (45) in the United States, and the Local Government Operations Research Unit (41) in the United Kingdom. Although these studies include some interesting features, they will not be considered here, as they are, in essence, developments of the studies already considered. De Donnea's study in Holland (16) contains interesting theoretical developments and will be discussed in Chapter 3.

One interesting approach neglected so far is that of Moses and Williamson (53). In some ways this work may be thought of as a development of the efforts discussed on pages 14-17, and is, in a sense, a transition between these efforts and the disaggregate, behavioral modelers. On the other hand, the explicit emphasis on estimating the diversion prices necessary to move commuters from the automobile to public transit sets this study in a category by itself.

Moses and Williamson distinguish two streams of thought in the valuation of time which correspond to the approaches outlined earlier; they are labeled, respectively, the "income" approach (which values time according to the worth of time in work), and the "pre-cost" approach (which values travel-time savings according to money-cost differentials between modes). From the point of view of this thesis, the pure-cost approach is the more interesting, since it approaches the valuation of time by way of a choice situation. When a traveler is observed to pay more (in operating costs, higher fare, or toll fees) to take a faster route or mode, his value of time can be inferred as being at least equal to the amount that he is willing to pay for a given time-saving. However, Moses and Williamson "are seeking . . . to determine a measure of the net benefit to the consumer, and thus must determine the price at which he would be indifferent between the two modes."[5] From this point of view, the pure-cost approach is unsatisfactory, since it ignores the wage rate; conversely, the income approach is equally unsatisfactory, since it ignores the time and cost characteristics of the modes. The rest of this paper is devoted to a synthesis of these two approaches and to the estimation of diversion prices.

It is not without interest to note that the authors refer to Warner's work as an interesting new development. Since this study is concerned with models of the Warner type, the work of Moses and Williamson will not be considered further.

Conclusion

It might appear that with the five recent behavioral studies mentioned above, the area of explaining choices and valuing time has been well covered. This is not the case. The explanation of commuting choices and the derivation of a value of time for commuters has been covered, but little is known about choices and values of time in other situations.

The primary aim of this study, therefore, is to extend the range of knowledge to include trips that are not commuting trips. The study will examine trips made between Edinburgh and Glasgow and their respective catchment areas, i.e., medium-range intercity trips. It should be remembered at this point that although some journey to work and business travel is included in the data collected, this study is concerned with travel in nonworking time.

2 The Development of the Model

This chapter develops formally a behavioral-choice model and shows how a value of time can be derived from such a model. The hypothesis that forms the basis of the model will be discussed, the necessary assumptions will be considered, and the form of the model will be examined. Finally, the derivation of a value of time will be demonstrated.

The Hypothesis

Since the aim of this chapter is to develop a model that will explain the modal choice of a group of travelers, the first step is to construct a hypothesis that will explain the mechanism by which a choice is made. While the exposition is couched in terms of a modal-choice model, the analysis applies equally to a route-choice model. A number of hypotheses could be advanced. For example:

Hypothesis 1: The traveler will choose the fastest mode.

Hypothesis 2: The traveler will choose the cheapest mode.

Hypothesis 3: The traveler will choose the most comfortable mode.

(It is acknowledged that other factors may be of importance. For simplicity of exposition the number has been restricted. Other factors will be introduced at a later point.)

Without doubt, each of these hypotheses will explain the modal choices made by some of the travelers, but they are all open to serious criticism. Each hypothesis claims that the traveler makes his modal-choice decision on the basis of information on one, and only one, aspect of travel: speed, cost, or comfort. In short, they lack generality. On the other hand, they all embody one noteworthy feature: they all hypothesize that choice is based on the *relative* features of the transport modes; the use of the term "fastest mode" implies relativity.

It might, however, be possible to combine these three hypotheses into a single, more general hypothesis:

Hypothesis 4: The traveler will choose the fastest, cheapest, and most comfortable mode.

This composite hypothesis has the advantage of a certain degree of generality, in that it includes a number of mode characteristics as factors influencing the traveler's choice of travel mode, but it has the obvious disadvantage of setting up a choice criterion that is virtually impossible to fulfill: it is unlikely, if

15

not impossible, for one mode to be "fastest, cheapest, and most comfortable." It thus becomes necessary to introduce notions of weights and trade-offs. The traveler will attach to each factor a weight that represents the relative importance of that factor to him. This will enable him to decide, given a mode choice in which the fastest mode is not the cheapest, the rate at which he is willing to trade off "cheapness" for "fastness" (speed).

Thus, the hypothesis can be reformulated in the following terms:

Hypothesis 5: The traveler will choose the mode whose speed, cost, and comfort advantages, weighted by their importance to him, produce the most favorable combination relative to the other modes.

This hypothesis explains the way in which a traveler considers the mode characteristics when making his choice of travel modes. It seems wrong, however, to stop at this point, considering only characteristics peculiar to the modes. Travelers themselves are a heterogeneous group with widely differing characteristics; moreover, they do not all make the same types of journey. It seems reasonable, therefore, to introduce into the hypothesis some conditions about the traveler and his type of journey.

Hypothesis 6: The traveler will choose the mode whose speed, cost, and comfort advantages, weighted by their importance to him, produce the most favorable combination relative to other modes. The weights will be in some part determined by the characteristics of the traveler himself, his environment, and the trip he is making.

This, then, is the hypothesis that forms the basis, explicitly or implicitly, of a behavioral mode-choice model. A similar hypothesis has been developed by the Local Government Operations Research Unit (44).

The Assumptions

It is now appropriate to examine the assumptions that are necessarily implied by the above hypothesis and to consider whether any further assumptions are necessary to allow us to proceed with our analysis of the problem. The first three assumptions are necessary to the hypothesis.

Assumption One: Rationality

"Rationality" means that a traveler will behave in the manner of the classical "economic man." He will try to maximize the satisfactions that he can obtain from his economic resources and will plan his actions to achieve this end. In simple terms, he prefers more of a good to less of it. This can be interpreted in the present context as meaning that he prefers a faster mode to a slower one, a cheaper mode to a more expensive one, etc. Thus, he may not "rationally" choose a mode that is both slower and more expensive *ceteris paribus* than the alternative. When other factors are added to the time and cost variables chosen to illustrate the exposition, the rationality must apply equally to them. Thus, the rational traveler will prefer more comfort, more security, and more convenience to lesser amounts of these characteristics.

Assumption Two: Limited Resources

In order for a meaningful choice problem to exist, it is necessary to assume that the economic resources available to the traveler are limited, i.e., his income is fixed. Should his resources be unlimited, he would be able to purchase speed or comfort regardless of cost. With fixed resources he must trade off these factors, the one against the other, in order to maximize his satisfaction.

Assumption Three: Perfect Knowledge

It is assumed that the traveler knows the characteristics of all the modes open to him; he knows the times, costs, levels of comfort, etc. of all the modes that fall within his range of choice. (It is possible that in certain situations, e.g., analysis of small time savings, the perfect knowledge assumption should be relaxed.)

These three assumptions are commonplace in the analysis of consumer demand, and it is thought that no objections will be raised to their use at this point. The following assumptions derive from the nature of a behavioral-choice model.

Assumption Four:
Preference Reflected in Behavior

It is assumed that the choice of a given mode by a traveler implies that he has considered the characteristics of the modes, relative to his own and his trip characteristics, and has reached the conclusion that one mode is preferable. As Little puts it: "We infer how a person feels from the way he acts."[1]

Assumption Five:
Two-Term Consistency[2]

This assumption serves to strengthen Assumptions One and Four in that it implies that preference can be inferred from single acts of choice. Briefly, it means that observations of choice behavior will not contradict each other. In other words, if the traveler chooses mode A when mode B is available, he cannot (consistently) choose mode B on another occasion when the same choice is open to him, and his personal characteristics have not changed. In an analysis of multimodal choices the assumption of transitivity would be useful. As this study is concerned with a binary choice, two-term consistency is sufficient.

These two assumptions form the basis of a behavioral-choice model. While it is acknowledged that cases may arise where a single act of choice may not reveal true preference, e.g. where the traveler is making a choice experimentally, it is contended that in the majority of cases the "single-act-revealed-preference" assumption[3] is not unreasonable.

The Form of the Model

The previous sections in this chapter have laid down the basis of the behavioral-choice model. This section seeks to establish a link between the hypothesis described above and a mathematical model that represents it. A mathematical model is no more than the representation of a hypothesis in terms of mathematical symbols, it being convenient to represent the hypothesis in such a way in order to employ statistical techniques to provide estimates of the weights attached to the variables and a measure of the significance (or strength) of the postulated relationship. In the case under discussion, the postulated relationship is between the choice of mode and a number of variables, such as relative times and costs, which are thought to influence that choice. In order to simplify the discussion and to allow diagrammatical representation without resort to three dimensions, the analysis will proceed in terms of a single independent (or explanatory) variable: the time difference between modes (ΔT).

A further simplification that can be introduced at this stage is that the analysis will be restricted to cases of binary choice, i.e., where only two modes are available. This restriction is introduced for two reasons. First, it greatly simplifies the analysis, and second, it can be argued that the choice of travel mode tends to reduce to a binary choice. Many travelers have little or no knowledge of more than two modes, and it would appear that the mode-choice decisions take the form of a hierarchy of binary choices. For example, a traveler, faced with a choice between bus, train, and car, might make a binary choice between the two public transport modes and then a further binary choice between the chosen public and the private transport mode (Figure 2-1).

Equally, a traveler faced with a bus, car, bicycle, or walk choice might first choose between the motorized or nonmotorized modes, and then choose the preferred mode within the selected pair. These two examples demonstrate the way in which a multimodal choice may be thought of in terms of a hierarchy of binary choices. To document the above assertions is a difficult task, but some indirect evidence may help. In the Edinburgh-Glasgow Area Modal Split Study,

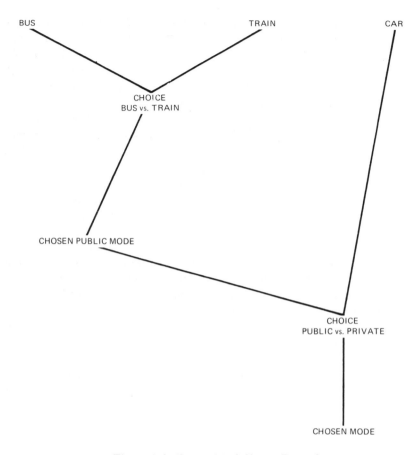

Figure 2-1. Hierarchical Choice Procedure

it was observed .that the questions most frequently not answered were those relating to the characteristics of the second-best mode. This may be interpreted as an indication that the subjects did not possess comprehensive information about that mode. If this interpretation is accepted, it is logical to assume that they will know less about the third-best and fourth-best modes. It is asserted, therefore, that the procedure may be construed as a hierarchy of binary choices which facilitates the decision-making process by reducing a multiple choice to a binary choice. It is possible that before the decision is made between the chosen and second-best modes, the lower-order choices may be highly speculative and based on habits and prejudices. Nevertheless, the final link in the chain is a binary choice. In order to ensure that the study examines the last choice in the hierarchy, the questionnaire must elicit information on the chosen mode and the second-best mode.

Similar recursive models have been used in housing studies where the "move/do not move" and the "buy/rent" choices have been considered hierarchically. Note that it is not intended to imply an explicit decision-making process; rather, an abstract construct is sought to facilitate the analysis of choices. The postulates leading to this construct are not unreasonable, and they facilitate both the data collection and analysis stages of a choice study.

Given, then, a choice between two modes, A and B, and a difference in time between mode A and mode B of ΔT minutes, it is now appropriate to consider the form of the relationship between choice and ΔT. (All other factors will be assumed to be equal for each mode.) One problem must be overcome, however, before even a simple linear relationship can be considered: the nature of the variable which represents choice. Since a binary-choice situation has been constructed, the choice variable can only take one of the two values. In this case, the choice variable can take values of zero or one, where zero and one represent the choice of mode B and A, respectively. The dependent variable can, therefore, be interpreted as the probability of choosing mode A, i.e., $P(A)$. Since the probability cannot be observed for an individual, the observed choice is used as a proxy dependent variable. The statistical problems associated with the use of this kind of variable will be discussed in Chapter 7.

Initially, the relationship will be given a simple linear form (Figure 2-2), and the discussion of the shortcomings of this form will demonstrate the rationale for choosing an alternative form.

That this form of the relationship is inappropriate is evident when its implications are considered. The fact that the equation passes through the origin implies that when ΔT is zero the probability of choosing mode A is zero, i.e., mode B will be chosen. This is not a practical result, since, *ceteris paribus*, the traveler should be indifferent between mode A and mode B when ΔT is zero. It

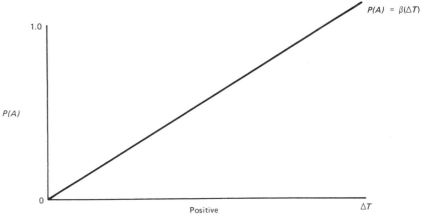

Figure 2-2. Basic Choice Relationship

is possible to correct for this by changing the scale of the ΔT axis. The result is shown in Figure 2-3.

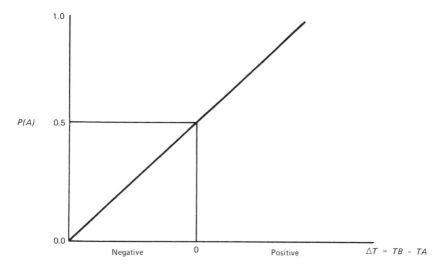

Figure 2-3. Choice Relationship–Modification 1

This form of the equation fulfills the requirement that the probability of choosing mode A should be 0.5 (i.e., the traveler should be indifferent) when ΔT is zero. The probability of choosing mode A will be unity when a certain positive threshold level of ΔT has been reached, and vice versa. The threshold levels can only be determined experimentally. This form is clearly an improvement over that shown in Figure 2-2, but the question arises as to whether a simple linear form, even suitably constrained, is satisfactory. The slope of the line representing the equation illustrates the rate at which the probability will change in response to a given change in ΔT, and clearly a straight line illustrates a constant rate of change $P(A)$ with respect to ΔT. It may be postulated, however, that the rate of change is not constant, but that as the threshold level of ΔT is approached, a given change in ΔT will lead to a smaller change in $P(A)$ than it would at lower absolute levels of ΔT. In other words, the curve will become flatter as it approaches the threshold levels of ΔT. Conversely, it is implied by this postulate that the curve will be steeper in the regions where ΔT is close to zero. Thus, the curve shown in Figure 2-4 results.

In economic terms, this may be interpreted as meaning that the marginal utility of increments of time difference will diminish as the time difference becomes larger. This result is intuitively appealing.

It should be noted at this point that different assumptions can be made about the form that the nonlinearity will take. For example, it could be assumed that

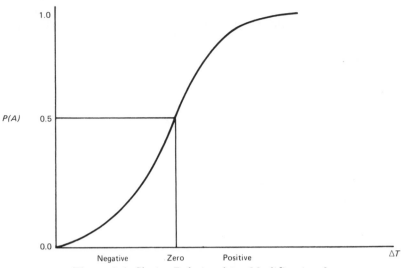

Figure 2-4. Choice Relationship—Modification 2

only large time differences influence choices and that small increments of ΔT have only a small effect, as shown in Figure 2-5.

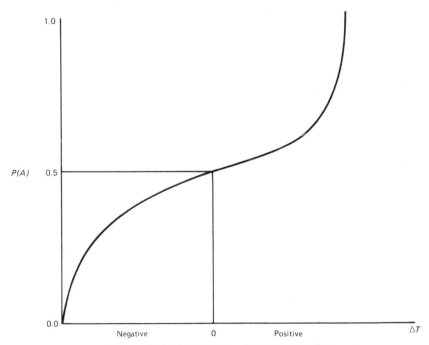

Figure 2-5. Modification 2—Alternate Form

It is argued, however, that the "S"-shaped curve developed above (Figure 2-4) is more plausible in terms of its economic interpretation.

The above argument has been developed in a somewhat indirect form in order to give greater weight to a more fundamental objection to both the linear form and the nonlinear form of Figure 2-5. The objection is that such formulations are unconstrained; in other words, they allow the dependent variable to take values greater than one or less than zero if a new observation is applied to the model (Figure 2-6). For example, a $\Delta T = -15$ observation would imply a negative probability, which is not possible. While it is recognized that all models may suffer distortions when the relationship is extended beyond the range of data on which it is based, it is contended that the extrapolations should not yield impossible results.

The sigmoid form avoids such problems and is based on a more realistic view of the rate at which $P(A)$ will change in response to changes in ΔT in different ranges of the curve: i.e., it is felt that a change in the range $\Delta T = 20 \rightarrow \Delta T = 25$ will be less important to the traveler than a change in the range $\Delta T = 5 \rightarrow \Delta T = 10$. In other words, the inclusion of diminishing returns is an important feature.

The form of the relationship shown in Figure 2-4 is the one considered in the following analysis. Methods of fitting such a curve to survey data and of estimating the coefficients are discussed in Chapter 7.

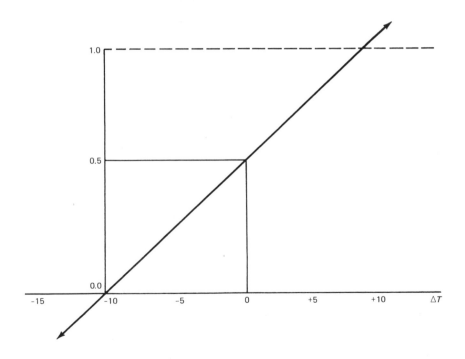

Figure 2-6. Extrapolation Problems of Unconstrained Forms

The Derivation of the Value of Time

Having discussed the development and form of the behavioral model that will be used to explain modal choices, it is now appropriate to turn to the other aspect of this study: the valuation of time spent traveling. This section will set out the mechanism by which a value of time can be derived from a behavioral modal-choice model.

In the above discussion the choice of mode was related to the difference in time between mode A and mode B, ΔT. It is now necessary to complicate the analysis slightly by introducing a further explanatory variable: the difference in cost between mode A and mode B, ΔC. It will be noted that the use of ΔT and ΔC at this point begs the question of which variables should be included in the relationship, and what form they should take. This matter will be discussed in Chapter 4. For the purposes of this exposition, it is assumed that the only relevant explanatory variables are ΔT and ΔC.

Let the modal-choice model take the form: $P(A) = f(\Delta T, \Delta C)$, and let the precise form be: $P(A) = a_0 + a_1 \Delta T + a_2 \Delta C$. a_0 is a constant term; a_1 and a_2 are the coefficients (weights) attached to the explanatory variables ΔT and ΔC, respectively. Note that this equation is not intended to represent the relationship in Figure 2-4. It is a simplification justified on the grounds of facilitating the exposition.

The value of time is defined as:

$$VOT = \frac{a_1}{a_2}$$

This ratio determines the amount by which, given a unit change in ΔC, ΔT must change in order to leave $P(A)$ unchanged. In other words, if ΔC changes by one unit, ΔT must change by $\frac{a_1}{a_2}$ units in order that $P(A)$ remain unchanged. If ΔC is in units of cents and ΔT in units of minutes, a one cent change in ΔC will be compensated for by $\frac{a_1}{a_2}$ minute change in ΔT. Therefore, the ratio $\frac{a_1}{a_2}$ represents the rate of substitution between time and cost, i.e., the value of time, since

$$a_1 = \frac{\partial P(A)}{\partial \Delta T} , a_2 = \frac{\partial P(A)}{\partial \Delta C} \quad \text{and}$$

$$\text{the value of time} = \frac{a_1}{a_2} = \frac{\partial \Delta T}{\partial \Delta C}$$

The derivation of a value of time outlined above is valid for any form of behavioral model which estimates coefficients for linear combinations of time and cost variables. It is applicable to the nonlinear forms of the model described above, as the probabilities are expressed as nonlinear functions of a linear combination of explanatory variables. For example, the logistic curve is expressed as:

$$P(A) = \frac{e^{(\alpha_0 + \alpha_1 X_1 + \alpha_2 X_2 + \ldots + \alpha_n X_n)}}{1 + e^{(\alpha_0 + \alpha_1 X_1 + \alpha_2 X_2 + \ldots + \alpha_n X_n)}}$$

The derivation of the value of time is not dependent on the form of the relationship, since its validity is based upon the logic of the mechanism by which time and cost are traded off in the making of a choice.

3

Behavioral Models and Economic Theory[1]

In Chapter 2 a hypothesis was developed to explain the mechanism by which travelers choose their mode of transportation. This chapter considers whether the behavioral models of mode choice based on this hypothesis have any basis in economic theory in general, and in the theory of consumer demand in particular. The first step will be to examine the work of previous analysts who have developed behavioral models of mode or route choice, in order to ascertain whether or not their models have any foundation in economic theory. This will be followed by a review of the results of some recent "allocation of time" work, which may throw some light on the question under consideration. Finally, a model of consumer behavior which explicitly includes travel will be developed, and some modifications of it will be discussed.

Justification for Behavioral Models

This section examines the work of the major behavioral modelers in order to discover the theoretical basis for their models. The analysts under consideration are Warner (82), Stopher (66 and 68), Quarmby (58), and Lisco (42), and it is interesting to note that, with the exception of Warner, who preceded the others, they carried out their work independently. Their work will be examined for any content that might lead to the conclusion that the models they estimate, and thus the hypotheses they test, have a basis in economic theory.

Warner's aim is to study the choices of mode made by individuals in varying circumstances. His objective is "the examination of the influence on consumer choice exerted by three economic variables,"[2] notably time, cost, and income. The study is set in stochastic terms; Warner is concerned with the estimation of the probability that a traveler will choose a given mode in response to certain characteristics of the alternative modes. The problem is treated in this way to allow quantification of the idea that changes in characteristics can lead to changes in choice behavior.

In fact, Warner does not attempt to justify his model beyond asserting that a binary choice exists and that this choice is influenced by a number of variables. While his treatment of the forms of the variables and of the statistical techniques that may be used to examine binary choices is meticulous, he provides no theoretical justification for his model and does not attempt to set his hypothesis in the context of a body of theory, economic or otherwise.

27

Stopher, in his study of a group of London commuters, takes a different approach. He chooses the influential characteristics from the results of a survey aimed at discovering which factors travelers considered most important when choosing their mode of transport. Four factors are important: time, cost, comfort, and convenience. Since the last two factors could not be quantified, the set of explanatory variables is automatically reduced to a subset of time and cost. Thus, Stopher chooses his variable set on pragmatic grounds, being largely concerned that his model should be correct in behavioral terms. In other words, he is anxious to ensure that the variables used are really those that travelers consider of importance.

Quarmby, on the other hand, is concerned with establishing his model on the basis of a meaningful theoretical development, which spells out the postulates and assumptions upon which the model depends. Initially, it is assumed that the traveler has made a decision to travel to a specific place for a specific purpose and that the problem is to develop a model that will describe the traveler's choice between the travel modes available to him. It is further assumed that there are k dimensions of travel, each of which gives rise to some "disutility"; the dimensions include travel time, walking time, travel cost, inconvenience, discomfort, etc. d_{pij} is defined as the measure of dimension p ($p = 1 \dots k$) for mode i ($i = 1 \dots h$) for person j ($j = 1 \dots n$). Thus, d_{pij} equal to ten minutes would be the value (or measure) of dimension p, say walking time, by mode i, say train, for the jth person. The importance of each dimension is indicated by a series of weights, λ_{pij}, which represent the contribution of each dimension to the total disutility, which is:

$$D_{ij} = \sum_{p=1}^{k} \lambda_{pij} d_{pij} \tag{3-1}$$

The traveler will then choose the mode that minimizes D_{ij}. Note that this formulation implies additivity of "utilities," i.e., the disutility associated with time is added to that associated with cost to obtain a sum of the disutilities associated with all the dimensions of the journey.

Quarmby alleges that all choices can be resolved into a choice between two alternatives and suggests that it is, therefore, relative disutilities which are of importance. In other words, it is not the absolute values of times, costs, etc. that influence choice, but a relative measure, such as a ratio or a difference. Given this postulate, it is no longer satisfactory to set up the simple minimization problem of minimizing relative disutility R_j, defined as:

$$R_j = D_{1j} - D_{2j} \tag{3-2}$$

It becomes necessary to set up a more general form of R_j to be minimized:

$$R_j = \sum_{p=1}^{k} \lambda_{pj} f_p \, (d_{p1j}, \, d_{p2j}) \qquad (3\text{--}3)$$

where f_p represents a more general way of expressing the relative measure of each dimension between modes. Thus, if:

$$X_{pj} = f_p(d_{p1j}, \, d_{p2j}) \qquad (3\text{--}4)$$

then

$$R_j = \sum_{p=1}^{k} \lambda_{pj} X_{pj} \qquad (3\text{--}5)$$

Given this formulation, the traveler then chooses mode 1 if R_j is low and mode 2 when R_j is high.

Quarmby at this point explains that

empirical work will examine which way of expressing "relativeness," i.e., which form of f_p comes nearest to explaining behavior. A more fundamental task will be to develop a method for finding the weighting factors λ, and for predicting what people will do when we know their R_j's.[3]

Thus, at this point he turns to a consideration of statistical techniques.

Although the introduction of the concept of utility into the development of his model might lead the reader to believe that Quarmby was attempting to derive the model in economic terms, this is not so. Quarmby uses the term "disutility" to represent the unpleasantness or costs of a journey, but he does not use the concept in its economic sense or in order to derive economic implications. His approach is closer to a cost-minimization approach, in which the traveler seeks to minimize the generalized costs of a given journey. Quarmby's major interest is not to derive his model in economic terms, but to demonstrate that the relative disutility model is the logical equivalent of a discriminant function.

Lisco's study is concerned with the journey to work from Skokie, a Chicago suburb, to the Loop, the central business district. It should be noted that the main aim of Lisco's study is "an attempt to put a value on the time spent by commuters during their daily travels to and from work."[4] The development of his model is, therefore, couched less directly in terms of explaining modal choices, but more in terms of deriving a value of time. The former is, however, an essential step in the latter. In an attempt to achieve the purest possible relationship between time and cost, Lisco develops his model in terms of a route rather than a modal choice, although his empirical work estimates a modal-choice model.

The background of the model is a situation in which commuters choose between a fast toll road and a slower free road, both roads being comparable in terms of quality and traffic. Lisco's model is based on the assertion: "If a given cost difference had the same effect upon commuters in their choice of highway as a given difference in time, then the two differences could be equated and a value of time determined."[5]

Assume that a group of commuters, faced with a choice between two routes identical in terms of time and cost, is split 50-50. Suppose then that a change occurs which makes one road faster, and that the split changes to 60-40 in favor of the faster road. Further suppose that the instigators of the improvement decide to levy a toll, the result of which is to reestablish the split at 50-50. It can be asserted, as Lisco does, that since the improvement and the toll have had the same effect on the behavior of the commuters, they can be said to be equal. If the improvement was five minutes and the toll was twenty-five cents, then the value of time would be determined as twenty-five cents for five minutes, or three dollars per hour.

Should factors other than time and cost be thought to influence choice of route, they must be either assumed to be equal for both routes or allowed for by the use of appropriate multivariate statistical techniques.

Other analysts, such as Lave (41) and the Local Government Operations Research Unit (44), have tended to accept the position developed by the first generation of modelers. (One exception is McGillivray (45), whose model was developed simultaneously with the one developed in the remainder of this chapter.)

It is appropriate to pause at this point to consider what has been gathered from the preceding accounts. In general, modal-choice modeling has been conceived of as the estimation of a relationship, of the form: Choice $= f$ [(Modal variables, $1 \ldots m$) (socioeconomic variables, $1 \ldots n$)]. The main problems with which modelers have concerned themselves have, therefore, been the forms and combinations of the dependent variables, and the statistical method to be used to estimate the coefficients.

If a number of models constructed on a pragmatic basis appear to reflect accurately the travel behavior of the study group, is it necessary to develop respectable economic antecedents for the model? The answer must be Yes, for two reasons. First, if the model is derived in terms of economic theory, one can be more confident about the relationships between the variables. Take, for example, the equilibrium condition derived from a utility function:

$$\frac{MU_i}{P_i} = \lambda \qquad\qquad \begin{aligned} MU &= \text{marginal utility} \\ P &= \text{price} \end{aligned}$$

From this simple condition it is possible to explain what happens to purchases of the *i*th good given changes in price. In a transport model that involves times, cost, comfort, etc., the use of a similar theory to derive the model may enable the analyst to follow the ramifications of the variables that result from any given change.

The second, and perhaps more important, reason is that unless the model is based firmly upon economic theory, it is difficult to know what it means. Some modelers, because they have not derived the theoretical implications of their models, are unsure of the meaning of their results and, indeed, of whether they are meaningful at all. Hence the arguments about whether values of time derived from modal-split models are "average" or "marginal" values, whether they are "general," "mode-specific," or "choice-specific." It is only through careful consideration of the theoretical implications of a model with a sound theoretical basis that meaningful interpretations of the results of the model can be made.

The Work of the Stanford Research Institute

Before concluding that the field is completely barren, it is important to consider the work of Haney (29) at the Stanford Research Institute. Since his research marked a major advance in the field, it will be considered in some detail.

In volume I of the study of commuters carried out by the S.R.I., Haney considers the problems the modeler meets when developing a mathematical model to explain route choices. The problems are:

1) to determine the factors that drivers consider in making highway decisions.
2) to develop a measure of the accuracy with which drivers are informed of those quantitative factors that are considered.
3) to assess the relation of the numerical value of a factor to its economic utility.
4) to determine ways in which individual utilities are combined into a total utility for each alternative considered.[6]

Haney begins with the assumption that the traveler is an "economic man." Although the concept of an "economic man" is commonplace in economic theory, its appearance in the derivation of a modal-split model was new and marked perhaps the first attempt to develop a model explicitly based on economic theory, with all the assumptions and implications spelt out. As far as the *Homo Oeconomicus* assumption is concerned, the implications are clear. The assertion that the traveler is completely informed means that he has knowledge of all the factors that enter into the model; he knows the journey times and costs by each alternative route, the toll fees, level of traffic, etc. Rationality means that he can rank alternatives and will obey the rules of transitivity: if route A is preferred to route B and route B is preferred to route C, then route A will be preferred to route C.

Haney also infers from the rationality assumption that the traveler follows some kind of maximizing process. In the terms of the theory of consumer choice, this means that the consumer chooses combinations of goods in order to maximize the utility that he obtains from these goods. In the route choice situation, the maximizing process can be restated as minimizing the total cost, in terms of time, money, and inconvenience costs of the given journey for a given trip purpose.

It is conceded that many travelers will not fulfill all the conditions of these assumptions. This does not mean, however, that the assumption should be abandoned. Lange says, "The postulate of rationality is justified only when the logical deductions agree with the result of empirical observation with an acceptable degree of approximation."[7] As the rationality assumption has formed the basis of a number of economic theories in the field of consumer choice, it seems reasonable to use it in the development of a route choice model.

Haney's next step is to introduce the concept of utility, and by way of the problem of combining utilities when interdependencies exist, the concept of indifference curves. If it is assumed that *ceteris paribus* time and cost are the influences on route choice, the traveler's indifference map can be represented as shown in Figure 3-1. It will be noted that the axes represent time and money used to fulfill a given trip and not, as is usual, amounts of goods consumed. This change has the effect of making the indifference curves concave to the origin and making the desirability of being on a given curve increase as the origin is

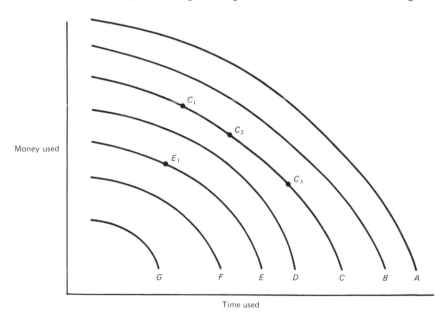

Figure 3-1. Haney's Indifference Map

approached. In a sense, it may be thought of more as a trip production possibility map.

The traveler whose indifference map is represented by Figure 3-1 will be indifferent between routes C_1, C_2, and C_3, since although they represent different expenditures of time and money, they all lie on the same indifference curve. He would, however, prefer E_1 to C_1, C_2, and C_3. It is theoretically possible, therefore, given a traveler's indifference map, to predict his route choice. Unfortunately, difficulties arise in producing an indifference map. While indifference maps have been produced experimentally, the process is expensive and time-consuming, since it involves subjecting a driver to a large number of hypothetical choice situations. Two other problems arise:

1. It is unlikely that choices are based on only two variables.
2. Maps for a large number of drivers would have to be derived before confident predictions about route choice in general could be made.

As an escape from this apparent impasse, Haney suggests a simplified form of indifference map derivation. Assuming that in real-life situations the difference between choices is relatively small, it is possible to divide the indifference map into sections; the procedure may be further simplified by assuming that the indifference curves in each section can be approximated by straight lines. Such a simplified indifference map is shown in Figure 3-2. For an indifference map of this type, "the relationships between variables can be written as linear equations."[8] In order to estimate a value of time for each section of the indifference curve, choices between alternatives lying within the section would be observed. The drivers choosing the time-saving alternative would have a value of time greater than that represented by the slope of the straight line connecting the two alternatives, and vice versa.

Haney concludes, "The indifference curve technique provides a theoretical basis for estimating relative trade-offs between varying amounts of different travel characteristics."[9] It would also "allow investigation . . . of the weightings which drivers appear to place on travel characteristics."[10] However, these weighting factors are only relative:

Determining appropriate numerical values for the weighting factors in a utility model may suffice to explain a decision process. . . However, development of values of time for economic justification of highway improvement alternatives requires that economic values be assigned to each of the important travel variables.[11]

It is contended that the behavior of a driver reflects the values that he attaches to each travel variable. The benefits to the driver of each course of action, in other words, choice, are the sum of the personal values associated with each

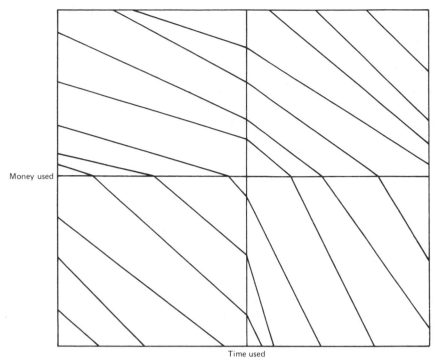

Money used

Time used

Figure 3-2. Straight Line Approximation of Indifference Map

variable. It is through a study of the driver's value systems that a value of time can be estimated.

Haney finally claims, "It appears that the best opportunities for placing economic values on time savings lie in the study of situations where there is a conscious payment of money and a conscious expenditure of time."[12] Estimation of the weighting factors is done by estimating the coefficients of a model that expresses the relationship between route choice and time and cost variables.

This attempt by Haney to justify the S.R.I. model explicitly in terms of economic theory marked a major advance in the field of modal- or route-choice modeling. The initial formulation of the choice decision in an indifference curve setting clearly explains the choice situation and the mechanism by which a choice is made. It is unfortunate that, owing to the problems of making indifference curve analysis operationally meaningful, this promising approach is difficult to develop. Nevertheless, Haney's work is an important step in the direction of economic justification, and his analysis paves the way towards demonstrating more conclusively that traveler mode choice should be analyzed in a setting of a trade-off between the times and costs of the alternative modes.

**Other Aspects of the Theory of
Consumer Behavior**

Since indifference curve analysis proved to be less than satisfactory, it was decided to reconsider some other aspects of the theory of consumer behavior. Let us examine the possibilities of deriving a utility model of choice which will form the basis of a behavioral route- or modal-choice model.

The first step in this examination is to consider some recent theories concerned with the allocation of time. These theories follow from the postulate that since the consumption of goods takes place in time, a theory that purports to explain choices should take account of not only the allocation of a fixed monetary budget, but also the allocation of a fixed time budget. A number of people have considered this problem, notably Becker (5), Johnson (36), and Evans (17); this work is summarized in a study carried out by Phillips (56) for the Ministry of Transport. For clarity of exposition, the discussion will proceed in terms of the work of Becker and Evans. This in no way implies that the work of the others mentioned above is less good. It is coincidental that a number of people produced similar work within a short period of time.

Becker's point of departure is the systematic incorporation of nonworking time into a traditional utility function. Existing theory claims that a utility function of the type

$$U = U(x_1, x_2 \ldots x_n) \tag{3-6}$$

is maximized subject to the resource constraint

$$p_i^1 x_i = I = V + W \tag{3-7}$$

x_i are market goods with prices, p_i^1, I is money income made up of earnings, W, and other income V.

It is assumed that instead of consuming market goods in their crude form, the consumer combines market goods with time to produce more basic, composite commodities, which are called Z_j:

$$Z_i = f_i(x_i, T_i) \tag{3-8}$$

Here x_i is a vector of market goods and T_i is the amount of time the consumer combines with x_i to produce Z_i. It will be noted that the consumer is thought of as both producing and consuming the Z_i. He combines time and goods via the "production functions," f_i, to produce Z_i, and chooses the best combination of Z_i by maximizing a utility function

$$U = U(Z_1 \ldots Z_m) \tag{3-9}$$

subject to a budget constraint

$$g(Z_1 \ldots Z_m) = Z \tag{3-10}$$

where g is an expenditure function of Z_i and Z is the bound on resources. "The basic aim of the analysis is to find measures of g and Z which facilitate the development of empirical implications."[1][3]

The procedure by which such measures can be derived involves maximizing the utility function (3-9) subject to separate constraints on the expenditure on market goods and on time and subject to the production functions in (3-8). The goods constraint can be written as

$$\sum_{i=1}^{m} p_i x_i = I = V + T_w \overline{w} \tag{3-11}$$

where p_i is a vector of unit prices for x_i; T_w is the hours spent at work and $\overline{w}$ is the wage rate vector. The time constraint is

$$\sum_{i=1}^{m} T_i = T_c = T - T_w \tag{3-12}$$

T_c is total time spent on consumption; T is total time. The production functions are

$$T_i = t_i Z_i \tag{3-13}$$

$$x_i = b_i Z_i \tag{3-14}$$

t_i is a vector giving the input of time per unit of Z_i, and b_i is a vector giving the input of market goods per unit of Z_i. In fact, the constrained maximization problem can be simplified when it is realized that the two constraints are not independent, since time is convertible into goods by using less time at consumption and more at work. Substituting for T_w in (3-11) gives

$$\sum p_i x_i + \sum T_i \overline{w} = V + T \overline{w} \tag{3-15}$$

and using (3-13) and (3-14), (3-15) can be rewritten as

$$\sum (p_i b_i + t_i \overline{w}) Z_i = V + T \overline{w} \tag{3-16}$$

This is a time and budget constraint, i.e., expenditure of time and money must equal the earning potential of the consumer. ($p_ib_i + t_i\overline{w}$) can be written as π and represents the full price of the unit of Z_i; it is equivalent to the prices of both the goods and the time used per unit of Z_i, i.e., the full price of consumption is equivalent to the sum of the direct and indirect prices.

Becker then shows that the resource constraint is only a meaningful construct if $\overline{w}$ is constant, i.e., if average earnings are constant. As this situation is unlikely, he suggests abandoning "the approach based on explicitly separate goods and time constraints and substituting one in which the total resource constraint necessarily equalled the maximum money income achievable, which will be called 'full income.' "[14]

It is not intended to deal with Becker's work at greater length as the subsequent work considered below leaves its findings open to question. The most interesting feature of Becker's analysis is that it represents a point of departure for attempts to introduce time costs into the choice process. His aim is the "systematic incorporation of non-working time" into the choice model, which leads to the finding that the full price of a unit of Z_i is the sum of the prices of both the goods and the time used per unit of Z_i. The use of two constraints, one for time and one for money, and the redefinition of consumption in terms of a Z_i, which might be called a consumption activity, are fundamental to further developments.

The exposition of the error that invalidates the generalized model of Becker is derived from Evans, although the ideas were promulgated almost simultaneously by Johnson and Oort (55), (q.v.). For expository purposes it is convenient to follow the arguments put forward by Evans.

The generalization of the traditional theory set out by Becker is incorrect because the utility function is misspecified. It is traditionally argued that the consumer maximizes a utility function of the form

$$U = U(L, G_i) \tag{3-17}[15]$$

L is leisure time and G_i are goods consumed. This formulation means that the consumer's satisfaction (utility) depends upon his income, represented by his ability to purchase goods, and the leisure time he has available for consuming these goods. This assumes that the number of hours worked by the consumer will not affect his utility, since an increase in working hours will decrease L but increase G_i. An implication of this assumption is the further assumption that the marginal utility of work is zero, i.e., that the consumer gains no utility and incurs no disutility through work, although he gains utility through leisure. Traditional theory involves a budget constraint

$$T_w\overline{w} = \sum G_iP_i \tag{3-18}$$

and a time constraint can be added:

$$T = L + T_w \tag{3-19}$$

$\overline{w}$ is the wage rate and T_w is the hours worked. T is total available time, and L is leisure (equivalent to T_c above). Following the argument set out by Becker, however, the two constraints can be combined into one, "because time can be converted into goods by using less time at consumption and more at work." Combining the two constraints gives

$$\sum G_i P_i = (T - L)\overline{w} \tag{3-20}$$

It will be realized that this is a parallel development to that of the previous section; the notation used by Evans has been simplified, and he takes no account of the "production function" aspects of Becker's theory. The simplified version is sufficient, however, to allow him to make his point as follows:

If (3-17) is maximized subject to (3-20), two first-order conditions result:

$$\frac{\partial V}{\partial L} = U_L - \lambda \overline{w} = 0 \tag{3-21}$$

$$\frac{\partial V}{\partial G_i} = U_{Gi} - \lambda p_i = 0 \tag{3-22}$$

λ is the Lagrangian multiplier associated with the budget constraint. At a maximum

$$\frac{-dG_i}{dL} = \frac{U_L}{U_{Gi}} \tag{3-23}$$

but if (3-21) and (3-22) are substituted into (3-23), the following result is obtained:

$$\frac{-dG_i}{dL} = \frac{\overline{w}}{p_i} \tag{3-24}$$

This means that the marginal rate of substitution of goods (income) for leisure is equal to the real wage rate. This result follows directly from the assumption outlined above that the marginal utility of work is equal to zero. The two constraints can be collapsed into one *only* because time is assumed to be freely convertible into money. This assumption is only valid because it is assumed that

$$U_W = 0 \tag{3-25}$$

and W is not an argument in the utility function.

It seems reasonable to question this assumption. It implies that people neither like nor dislike their work, that they get neither utility nor disutility from it. Evans gives the example of footballers who if they are amateurs may enjoy the game, but if they are professionals must derive no utility from playing. However, what is a pastime to one person may be a job to another: a bricklayer may do political work in his spare time, while a politician may build a wall. It is clearly unjustifiable to claim that the person carrying out an activity in his leisure time may obtain utility from it, whereas a person who carries out the same activity during working hours may not.

The next step in the argument is to incorporate hours worked into the utility function, thus

$$U = U(W, L, G_i) \tag{3-26}$$

The constraints are

$$T = L - T_w \tag{3-27}$$

and

$$\sum G_i p_i = T_w \bar{w}$$

The two constraints cannot be combined in this formulation, since it is no longer assumed that time can be freely converted into money through work without altering the consumer's utility level. Maximizing subject to both constraints yields:

$$\frac{\partial V}{\partial L} = U_L - \mu = 0 \tag{3-28}$$

$$\frac{\partial V}{\partial W} = U_W + \lambda \bar{w} - \mu = 0 \tag{3-29}$$

$$\frac{\partial V}{\partial G_i} = U_{Gi} - \lambda p_i = 0 \tag{3-30}$$

As Evans rightly points out, the most interesting condition is (3-29). This means that

$$U_W = 0 \tag{3-31}$$

only in the special case where the marginal utility of the wage rate just compensates for the marginal utility of the leisure foregone, i.e., where

$$\lambda \overline{w} = \mu \qquad (3\text{-}32)$$

The interpretation of these findings is clear and in accordance with observable evidence. When

$$U_W < 0 \qquad (3\text{-}33)$$

there is some element of disutility in the work, which means that $\lambda \overline{w}$ must be higher than is necessary to compensate for foregone leisure alone, since it must compensate also for the disutility in the work. On the other hand, when

$$U_W > 0 \qquad (3\text{-}34)$$

then $\lambda \overline{w}$ is not as high as would be necessary to compensate for foregone leisure, since part of the utility of the leisure foregone is compensated for by the utility inherent in the work.

It will be clear that this formulation of the theory allows the traditional situation where

$$U_W = 0; \qquad (3\text{-}35)$$

It is not necessary, however, that this condition hold. This result would seem reasonable, since there is no reason why the marginal utility of work should not vary, being positive in some occupations, negative in others, and zero in yet others. It can be claimed, with some justification, that the reformulated theory allows more to be explained than the traditional version, and is, therefore, on the principle of *Occam's razor*, to be preferred.

Having reformulated the traditional theory in more acceptable terms, the next step is to generalize the model. Evans constructs three categories of leisure activity, representing different combinations of time and money costs; some goods require time for consumption but are free, whereas others require both time and money expenditures, etc. This distinction is intended to demonstrate that the price of consuming a good is made up of money and time components, and leads into the redefinition of all three categories of leisure as "activities," a_i, which require inputs of either time or money, or both.

The new utility function is

$$U = U(a_i) \qquad i = 1 \ldots n \qquad (3\text{-}36)$$

a_i being the amount of time spent on the ith activity.

The utility function is maximized subject to the two constraints

$$K = \sum_{i=1}^{n} a_i \qquad (3\text{-}37)$$

and

$$\sum_{i=1}^{n} r_i a_i = 0 \qquad (3\text{-}38)$$

r_i is positive if the individual pays for the activity, negative if he is paid (work), and zero if the activity is free, and is expressed in terms of an hourly rate.

The following first-order equilibrium condition results

$$\frac{\partial V}{\partial a_i} = U_{ai} - \mu - \lambda r_i = 0 \qquad (3\text{-}39)$$

or

$$U_{ai} = \mu + \lambda r_i \qquad (3\text{-}40)$$

This means that, at the margin, an activity will be undertaken when the marginal utility of the activity just compensates for the marginal utility of the leisure time foregone and the marginal utility of the cost of undertaking the activity. If the activity is work, the r_i will be negative and may or may not swamp the positive μ, to make U_{a_i} negative. The marginal rate of substitution of one activity for another, M.R.S., is

$$\text{M.R.S.} = \frac{-da_j}{da_i} = \frac{U_{ai}}{U_{aj}} = \frac{\mu + \lambda r_i}{\mu + \lambda r_j} \qquad (j \neq i; \ r \neq 0) \quad (3\text{-}41)$$

As a result of the introduction of the time constraint, the M.R.S. is not equal to the price ratio, as it is in the traditional model. The Lagrangian multiplier associated with the time constraint, μ, modifies the M.R.S. Thus, (3-40) can be rewritten as

$$U_{ai} - \mu = \lambda r_i \qquad (3\text{-}42)$$

from which it follows that

$$\frac{U_{ai} - \mu}{U_{aj} - \mu} = \frac{\lambda r_i}{\lambda r_j} \qquad (3\text{-}43)$$

or more usually

$$\frac{U_{ai} - \mu}{r_i} = \frac{U_{aj} - \mu}{r_j} = \lambda \qquad (3\text{--}44)$$

This means that the consumer allocates time to his n activities in such a way that the ratio of the marginal utility of the activity net of the marginal utility of time to the cost of the activity is constant and equal to the marginal utility of money. He will not allocate time to any activities whose U_{a_i} is greater than μ, and the more U_{a_i} exceeds μ, the more he will be prepared to pay. (In the case of work, U_{a_i} could be less than μ, but this would be compensated for by the fact that r_i would be negative, leaving the ratio positive and equal to λ.)

It is important to note that it is not the marginal utility of the activity, net of μ, which will be the same for all activities, but the ratio of the net marginal utility to the cost of the activity. It should be obvious that the marginal utility to be derived from various activities can and will differ. Such differences are allowed for in the differing prices.

It is appropriate at this point to pause and consider the position of the analysis. Based on the work of Becker and Evans, a theory of the allocation of time has been developed which explains how a consumer will allocate his fixed time resources to a number of activities, basing his choice on the marginal utilities to be gained from the said activities and on their costs. It remains to consider what has been learned from the above consideration of allocation of time models which may be of use in developing a theoretical justification for behavioral models of mode choice. Although the allocation of time models are not directly applicable to this end, two features are noteworthy and must be kept in mind during the subsequent developments. The first interesting feature is the concept of a composite entity, an activity, which embodies both the time and commodity inputs involved in a given use of leisure time. This concept will be utilized later in the modification of a model that has travel time identified as an argument in the utility function. The second interesting feature involves the attempts to aggregate the time and money costs of an activity, when they are expressed in different units. The discussions of this should serve as a warning in the developments that follow.

A Utility Model with Travel Time Identified

The above discussion of allocation of time theories has diverted the discussion somewhat from its original course. It is now appropriate to return to that course by considering explicitly the role of travel time in a model of consumer behavior. In this model transport time is introduced as an argument in the utility function in an attempt to investigate the way in which the consumer undertakes his budgetary allocations between different goods with different amounts of travel attached to them. Transport time is brought into the discussion at this point because it is unrealistic to consider goods in isolation, since many goods have travel attached to their consumption and the choice of goods by the consumer may well depend to some extent on the amount of travel involved.

Consider the situation where the consumer has to choose between seeing an average film at a neighborhood theater, with only a short walk involved, and seeing a good film at a distant theater, with a car journey involved. It is postulated that he will only choose the distant film if the extra utility from that film compensates for the extra generalized cost of the travel involved. Such a situation provides the rationale for the introduction of travel time as an argument in the utility function, which becomes

$$U = U(L, T_w, T_i, G_i) \qquad (3\text{--}45)$$

L is the leisure time, $T_{\hat{w}}$ is work time, T_i is the travel time associated with the consumption of the good G_i. This function is to be maximized subject to the budget and time constraints.

$$\sum G_i p_i + \sum T_i c_i = T_w \overline{w} \qquad (3\text{--}46)$$

$$L + T_w + \sum T_i = T \qquad (3\text{--}47)$$

P_i is the price of the ith good, c_i is the cost per unit of travel associated with consumption of the ith good, $\overline{w}$ is the wage rate, and T is constant. The four first-order conditions for an utility maximum are:

$$\frac{\partial V}{\partial L} = U_L - \mu = 0 \qquad (3\text{--}48)$$

The marginal utility of leisure is constant and equal to μ [see Evans, (3-28)].

$$\frac{\partial V}{\partial W} = U_W - \mu - \lambda \overline{w} = 0 \qquad (3\text{--}49)$$

This means that the marginal utility of work is equal to the marginal utility of leisure less the marginal utility of the wage rate and is equivalent to Evans, (3-29)

$$\frac{\partial V}{\partial G_i} = U_{G_i} - \lambda p_i = 0 \qquad (3\text{--}50)$$

This means that the ratio of the marginal utilities and prices of the i goods is constant and equal to λ, the marginal utility of money [see Evans, (3-30)]

$$\frac{\partial V}{\partial T_i} = U_{Ti} - \mu - \lambda c_i = 0 \tag{3-51}$$

This condition is new and results from the inclusion of travel time as an argument in the utility function. It means that the marginal utility of time spent traveling is equal to the marginal utility of leisure time plus the marginal utility of the cost of traveling. In other words, the utility obtained from travel must, at the margin, be equal to the utility of the leisure foregone plus that of the cost of traveling.

Since the marginal utility of travel time has now been isolated in the model, it is possible to introduce a behavioral concept into the explanation of the consumer's allocation of resources. It will be clear from the introductory remarks of this section that some travel time has been associated with the consumption of each good. As consumption of most goods involves travel of some sort, it is contended that this formulation is reasonable. Should the "travel" be merely a short walk, then T_i will be very small and c_i will be zero; in the event that consumption requires no travel at all, both T_i and c_i will be zero.

It is now postulated that *consumption decisions are made on the basis of the utility derived from consuming a good, its price, the utility of the travel time associated with that consumption, and the price of the travel.*

If this is so, the traveler considers both U_{G_i} and U_{T_i}. From (3-50) and (3-51)

$$U_{G_i} + U_{T_i} = \lambda p_i + \mu + \lambda c_i \tag{3-52}$$

thus

$$\frac{U_{G_i} + U_{T_i} - \mu}{p_i + c_i} = \lambda \tag{3.53}$$

This result is an expanded version of the standard proportionality result of traditional theory and of (3-44), a modification of Evans' result. It means that, in equilibrium, the consumer will allocate his resources in such a way that he will only undertake activities for which the ratio of the marginal utility of the good, net of the marginal utilities of both travel and leisure time, to the cost of the activity (including travel cost) is constant and equal to λ, the marginal utility of money.

The Meaning of μ

The model developed above explains consumer behavior in terms of the relationship between the marginal utilities of goods and travel, their respective costs, and μ. Thus, μ is interpreted as the marginal utility of leisure time as opposed to that of work or travel time. Analysis of the composite equilibrium condition of the "transport identified" model

$$\frac{U_{G_i} + U_{T_i} - \mu}{p_i + c_i} = \lambda \qquad (3.54)$$

reveals that the consumer allocates his resources to the consumption of different goods on the basis of the net utilities. The interpretation of

$$U_{G_i} + U_{T_i} \qquad (3\text{-}55)$$

presents no problems: it is the marginal utility of consuming the ith good net of the marginal utility of the travel associated with the consumption of good i.

The next step is to subtract μ, the marginal utility of leisure, but the question arises as to what is meant by the marginal utility of leisure. Clearly it is a constant, and therefore it cannot be interpreted as the marginal utility of an activity carried out in leisure time, since different activities would have different marginal utilities and would not be constant. The solution of the problem can be approached through consideration of a difficulty associated with the "transport identified" model: leisure time and goods and services consumed in leisure time cannot be separated. μ is the Lagrangian multiplier associated with L in isolation and represents the marginal utility of leisure time in its purest sense, that is, leisure time in which no consumption takes place—activity-free leisure. Evans confirms this notion in his discussion of the "category-mistake" committed by Johnson and Oort.[16] To commit a "category-mistake" is to represent something as belonging to one logical type or category, when in fact it belongs to another; in this context, it means to confuse the value of time used on a particular activity with the value of time in general.

The price (value) of time in any activity will depend upon the activity and in many circumstances it will be equal to zero. But the marginal valuation of time in general for the consumer is the same whatever the activities he is engaged in.[17]

In other words, the value of time in a pure, activity-free sense becomes confused with the value of time used to pursue an activity. Then, μ can be interpreted as the value of pure, activity-free time, i.e., the value of time spent doing absolutely nothing. This being the case, it means that the concept of the marginal utility of leisure (μ) is purely abstract; time is a resource that can only be spent in doing something, and μ is the value of time spent doing nothing. It has been argued that μ cannot be the value of doing absolutely nothing, since doing absolutely nothing is impossible. Interpreted as an abstract concept, however, μ need not be an operational concept; it need not represent something performed by a human being. The abstract nature of μ is confirmed if the analogy with λ is considered, for λ can be interpreted as the marginal utility of money, in the sense that it is the value of relaxing the budget constraint by one unit. In the same way, μ can be regarded as the value of relaxing the time constraint by one unit. In other words, μ is the value of an increase by one unit in the total time available. Since the total available time, unlike the budget, is physically fixed, an increment to the day can only be considered as a hypothetical possibility.

An alternative way of interpreting μ involves a development of the abstract aspects of μ. Thus, it is possible to contend that time has no intrinsic value, but only carries an economic rent as a result of its scarcity. Since the number of hours in each day is fixed, and a certain number of these hours must be devoted to sleeping, maintaining life system, and working, the amount of time available for leisure activities is limited. The result is that activities that require time bear a time cost because the amount of time available is not unrestricted. Since competing activities require inputs of the limited amount of time, the available time must be rationed. This is effected by the device of attaching a time cost as well as a money cost to activities.

In conclusion, it can be said that μ is a concept of little operational significance. Its effect is to make the

$$U_{G_i} + U_{T_i} - \mu \qquad (3\text{--}56)$$

measure into a measure in which all values are expressed in terms of differences from μ. It is clear that μ cannot be interpreted as the marginal utility of leisure time in general, and, in particular, it cannot be interpreted as the marginal utility of time spent traveling. Thus, whatever interpretation is placed upon the values of time derived from behavioral modal-split models, they cannot be thought of as estimations of μ.

While the model developed above is interesting, both in the sense that it parallels traditional results while introducing travel time, and in the sense that it permits the interpretation of the "value of leisure time" to be clarified, it does not completely solve the problems outlined earlier in this chapter. This model does not explain the mechanism by which a consumer allocates his limited time and money resources to the consumption of a variety of goods. In this sense, it does not follow the line of development of a model from a traditional model through Becker to Evans which was set out in this chapter. It will be remembered, however, that the aim of this chapter was to provide an economic background or justification for the behavioral-choice model. Allocation of time theories were introduced only as an aid to fulfilling the main aim of the chapter. The development of the model has led away from allocation of time theories and towards one more disaggregate in the sense that it distinguishes different types of goods. It is justified if it leads to an acceptable rationale for the behavioral choice model.

One objection to the model developed above is that it fails to take into account both the time costs of consumption and the inputs to travel other than time. This dichotomy results from the fact that goods and traveling are not treated in the same way. In fact, both the leisure activities themselves and the travel activities associated with them are composed of inputs of consumption goods and time. The modifications of this model set out below take account of this deficiency.

A further objection to this model lies in the specification of the utility function and is concerned with the juxtaposition of leisure (L) and goods (G_i) as arguments in the utility function, thus implying that leisure may be regarded as distinct from the goods consumed and, hence, from any travel associated with that consumption. The true position is that once leisure time has been determined by the specification of working time, the consumption and travel activities take place in leisure time. In the modification of this model presented below an attempt to solve this problem is made by assuming that the work/leisure time distinction is given, so that the model deals only with activities carried out in leisure time.

The final objection has been regarded as the most serious objection to the model and arises from the last first-order equilibrium condition

$$U_{T_i} + \mu + \lambda c_i \tag{3-57}$$

As both μ and λ are constants, and since c_i is always positive, it is not possible for U_{T_i} to be negative. This means that the utility of time spent traveling cannot be negative, i.e., travel cannot give rise to disutility. This finding is consistent with the position that the consumer will only undertake activities, or consumption, which yield him utility. Thus, for a person to travel, he must derive utility from the travel. This position, however, destroys the interpretation of behavior which explains behavior in terms of the utility of the activity net of the utility (assumed negative) of the travel associated with it.

It is clear that there is a confusion between utility as envisaged by writers like Quarmby and as envisaged by an economist. The confusion is between unpleasantness and disutility. For example, a visit to a dentist may be unpleasant, but the sufferer derives utility from it. It is argued that transport falls into the same category. (That some people may desire pleasure, as well as utility, from travel is taken care of in the formulation developed in the next section.) The problem, therefore, is one of dealing with the dichotomous nature of certain activities that, like transport, both yield utility and are, to a certain extent, unpleasant.

Modifications to the "Travel Time Identified" Model

As many of the problems encountered in the development and interpretation of this model stem from an inability to distinguish between disutility and unpleasantness, this section will develop two new approaches that lead to a solution to the problem. The first is derived from the recent work of de Donnea[18] and introduces the distinction between the utility derived from the use to which time is put and the circumstances under which the time is spent; the second involves modifications to the models described earlier in this chapter, which produce results very similar to those of de Donnea's.

The innovative feature of de Donnea's work is his treatment of the problem of the dichotomous nature of utility; he seeks to distinguish between the utility of travel which is positive *ex hypothesi* and the so-called disutility associated with traveling. The former is positive, because travel enables the traveler to undertake activities that he could not undertake at his base location; the latter is negative, because travel involves a certain amount of inconvenience or discomfort.

In general terms, de Donnea considers

Time spent in some consumption activities may have a large positive utility, because it is an essential input in the production of a useful activity, but this time may simultaneously produce some kind of dissatisfaction for the consumer because it must be spent in disagreeable circumstances, for example, the time spent at the dentist's or in crowded trains on the journey to work.[19]

Thus, de Donnea works with the utility function

$$U = U[A_i, L(t_i)] \tag{3-58}$$

where A_i is the level of the ith activity, ($i = 1 \ldots m$), and $A_i = f(X_{ki}, t_i)$, ($k = 1 \ldots n$). $L(t_i)$ = the satisfaction or dissatisfaction resulting from the circumstances under which the time is spent. This initial model is developed to include travel services and time explicitly in the utility function, and this formulation is most relevant to the discussion at hand.

The utility function is

$$U = U(A_i, t_i, t_i^*, T_w) \tag{3-59}$$

A_i = level of the ith activity

t_i = time associated with the production of the ith activity

t_i^* = travel time associated with the ith activity

$T_{\hat{w}}$ = work time

and

$$A_i = f(X_{ki}, t_i, X_{ki}^*, t_i^*) \tag{3-60}$$

X_{ki} = amount of the kth good used in the ith activity

$$X^*_{ki} = \text{amount of the } k\text{th good used in the travel associated with the } i\text{th activity}$$

Thus, the two aspects of time, consumption time and travel time, enter into the utility function twice: once through A_i when they represent the positive utility-producing aspect of time used in the activity, and once by themselves when they represent the utility, which may be positive or negative, resulting from the circumstances under which the time is spent. Thus, t^*_i may yield positive utility, because it enables A_i to be undertaken, and negative utility, because it must be spent on a crowded train. This utility function is maximized subject to two constraints:

$$\sum_{k=1}^{n} \sum_{i=1}^{m} p_k x_{ki} + \sum_{k=1}^{n} \sum_{i=1}^{m} p_k x^*_{ki} = \overline{w} T_w \tag{3-61}$$

$$\overline{w} = \text{the wage rate}$$

and

$$T = \sum_{i=1}^{m} t_i + \sum_{i=1}^{m} t^*_i + T_w \tag{3-62}$$

$$T = \text{Total time}$$

and yields the following first-order conditions for a utility maximum:

$$\frac{\partial U}{\partial A_i} \cdot \frac{\partial A_i}{\partial X_{ki}} = \lambda p_k \tag{3-63}$$

$$\frac{\partial U}{\partial A_i} \cdot \frac{\partial A_i}{\partial X^*_{ki}} = \lambda p_k \tag{3-64}$$

$$\frac{\partial U}{\partial A_i} \cdot \frac{\partial A_i}{\partial t_i} + \frac{\partial U}{\partial t_i} = \mu \tag{3-65}$$

$$\frac{\partial U}{\partial A_i} \cdot \frac{\partial A_i}{\partial t^*_i} + \frac{\partial U}{\partial t^*_i} = \mu \tag{3-66}$$

$$\frac{\partial U}{\partial T_w} + \lambda \overline{w} = \mu \tag{3-67}$$

The first two conditions are standard results, but they may be combined to show that the marginal utility of good k must be equal in all its uses, whether as in input to the activity itself or to the travel process; the fifth condition is also a standard result. The third and fourth conditions, however, are more interesting.

Setting

$$\frac{\partial U}{\partial t_i} = 1_i \tag{3-68}$$

which is interpreted as the marginal utility arising from the circumstances under which t_i is spent, and

$$\frac{\partial U}{\partial t_i^*} = 1_i^* \tag{3-69}$$

which is interpreted analogously, we can write

$$\frac{\partial U}{\partial A_i} \cdot \frac{\partial A_i}{\partial t_i} = \mu - 1_i = \pi_i \tag{3-70}$$

where π_i is the marginal utility of the time used as an input in A_i, and

$$\frac{\partial U}{\partial A_i} \cdot \frac{\partial A_i}{\partial t_i^*} = \mu - 1_i^* = \pi_i^* \tag{3-71}$$

where π_i^* is the marginal utility of the time used in the travel associated with A_i. Then, from (3-65) and (3-66)

$$\mu = \pi_i + 1_i = \pi_i^* + 1_i^* \tag{3-72}$$

In other words, the marginal utility of time μ is the same for all consumption and travel activities at the margin, but it may be decomposed into two components: the marginal utilities of the time as an input and of the circumstances under which the time must be spent. Thus, the combined marginal utility of travel time will be equal to the marginal utility of the time needed to produce the activity itself.

This conclusion can be illustrated by an example. Imagine an individual who wants to spend a Sunday afternoon on the beach. It is realistic to assume that the farther the beaches are from the metropolitan area, the more agreeable they will be (cleaner, less crowded, etc.). Hence the farther the individual will travel, the more pleasant his time on the beach will be, but also the shorter this time will be. The individual will increase his travel distance as long as the marginal utility of his travel time is larger than the marginal utility of the time spent on the beach . . . at the distance already travelled.[26]

From (3-64) and (3-69)

$$\frac{\dfrac{\partial U}{\partial A_i} \cdot \dfrac{\partial A_i}{\partial X_{ti}^*}}{\dfrac{\partial U}{\partial A_i} \cdot \dfrac{\partial A_i}{\partial t_i^*}} = \frac{\lambda p_k}{\pi_i^*} = \frac{p_k}{\pi_i^*/\lambda} \tag{3-73}$$

But

$$\frac{\dfrac{\partial U}{\partial A_i} \cdot \dfrac{\partial A_i}{\partial X_{ki}^*}}{\dfrac{\partial U}{\partial A_i} \cdot \dfrac{\partial A_i}{\partial t_i^*}} = \frac{- dt_i^*}{dX_{ki}^*} \tag{3-74}$$

which is the marginal rate of technical substitution between the time and the good k used in the production of the travel associated with the activity, A_i. Thus

$$- \frac{dt_i^*}{dX_{ki}^*} = \frac{p_k}{\pi_i^*/\lambda} \tag{3-75}$$

i.e., the marginal rate of substitution between the time and the good k used in travel is equal to the ratio of their prices.

This result of de Donnea's provides a basis for using trade-off models to represent mode-choice decisions and to investigate the value of time, since if the price of the good and the price of time are constant (which they are assumed to be for a given trip), goods and time can only be substituted for each other by giving up one as the other increases. The cost of travel can be thought of as command over the goods used in travel, in the same way as a utility function may contain as arguments either goods themselves, or income, which represents command over goods. Then, it is logical to examine the trade-off between time and cost as the basis for models of mode-choice behavior.

An Alternative Formulation

The exposition of de Donnea is based on an innovation which may be regarded as rather cumbersome: the concept of utility deriving from the circumstances under which travel is spent. The alternative formulation presented in this section is based on a simplification of the problem. It will be remembered that the object at this point is to provide a theoretical basis for behavioral models of mode choice. However, mode-choice models are conventionally used as part of a system of models, often referred to as the U.T.P. (Urban Transportation Planning) Package, and comprising a trip generation model, a trip distribution model, a modal-split model, and a traffic assignment model. Each model takes the output of the previous models as given, so that the mode-choice model starts from a point where origins and destinations are given. This situation has been accepted by the disaggregate mode-choice modelers such as Quarmby, Stopher, and Lisco. As the mode-choice model takes as given the origins, the destinations, and, by inference, the activities, the objectives of this chapter can be met by considering a more restrictive formulation than those employed above.

Thus, we assume that the consumer has to decide between competing modes of transport which, for simplicity, we assume to be located on a continuum of alternatives. Since the origin and destination of the trip are fixed, the utility derived from the trip will be the same whichever mode is chosen (ignoring de Donnea effects), and the problem can be converted to one of producing a given product at minimum cost, i.e., of minimizing the cost, in money and time, of producing a trip.

If the trip production function is

$$T_i = f(X_{ij}, t_i)$$

X_{ij} = the goods and services used to produce the trip T_i

t_i = the time used

(3–76)

where T_i is the given trip, form the function:

$$Z = \sum X_{ij} p_j + \sum t_i V + \lambda[T_i - f(X_{ij}, t_i)] \qquad (3\text{–}77)$$

p_j = prices of the goods X_j

V = the value of time

(V is introduced here as the value (or price) of time for convenience in the derivation.)

Then, the first-order conditions for a minimum-cost solution are:

$$\frac{\partial Z}{\partial X_{ij}} = p_j - \lambda f'(X_{ij}) = 0 \qquad (3\text{–}78)$$

and

$$\frac{\partial Z}{\partial t_i} = V - \lambda f'(t_i) = 0 \qquad (3\text{–}79)$$

Dividing the former condition by the latter to eliminate λ gives:

$$\frac{p_j}{V} = \frac{f'(X_{ij})}{f'(t_i)} \qquad (3\text{–}80)$$

i.e., the ratio of the marginal products of the goods and time used to produce the trip, T_i, must be equal to the ratio of their prices. But

$$\frac{f'(X_{ij})}{f'(t_i)} = \frac{-dt_i}{dX_{ij}} \qquad (3\text{-}81)$$

In other words, in order to minimize the cost of a given trip, the traveler must equate the marginal rate of substitution between goods and time to the ratio of their prices. This is analogous to the result of de Donnea in (3-75).

The implication of this result is that the traveler will trade off time against goods (in other words, the cost) until he reaches the minimum-cost situation. This result, therefore, provides a theoretical basis for the use of behavioral disaggregate models in research into mode choices.

Conclusions

The above examination has shown that behavioral models of mode choice are not inconsistent with the theory of consumer demand. The development of this basic theory, together with inputs from allocation of time theories, lead to an initial, crude model of traveler behavior. Consideration of the objections to the crude model lead to improved formulations that demonstrate that the problem can be considered either as a modified utility-maximization problem or as a cost-minimization problem. In either case, the necessary conditions for either a utility maximum or a cost minimum imply that the traveler must equate the marginal rate of substitution between the time and goods used in the trip to the ratio of their prices. It may be inferred that this result will be achieved by trading off time against goods. Thus, the model based on a hypothesized trade-off between time and cost (or command over the input of goods) is shown to be consistent with the theory of the consumer (and of the firm) and to have a basis in economic theory.

4 Variables and Variable Forms

This chapter investigates the range of variables available for use in explaining modal choice and selects those variables that appear most promising from the point of view of the intercity modal-choice situation. The main source of guidance in such an investigation is the findings of previous studies in the field, although judgment must be exercised to ensure that the choice of variables clearly reflects the new choice situation under consideration. Table 4-1 shows the main variables used in the most important studies in the field. The form of the variable is not noted in this table, since the procedure to be adopted involves the discussion of each variable in turn. Therefore, in the following sections the different forms of each variable will be examined. The review of potential variables and variable forms will not assume a rigid format, and discussion of new variables or new variable forms will be introduced where appropriate. The extent to which data subdivisions might be useful will be examined as the variables concerned enter the review.

Table 4-1
Variables Used in Previous Studies

Variable	Lave	Lisco	Quarmby	Stopher	Thomas	Warner
Time	*	*	*	*	*	*
Cost	*	*	*	*	*	*
Comfort/Convenience	*				*	
Distance	†				*	*
Journey Purpose	n.a.	n.a.	n.a.	n.a.	n.a.	#
Journey Frequency						
Age/Sex	*	*			sex only	*
Demand for Availability of Car		*	*			*
Car Ownership	n.a.	n.a.			n.a.	n.a.
Size of Traveling Party						
Income	†				*	*
Other		*			*	

* indicates used as a variable
\# indicates used to stratify sample
† indicates used indirectly

Review

This section does not report in detail the forms of the variables used by analysts in previous studies. Rather it presents a consensus of which forms are potentially useful. Three basic types of variables are discussed: variables concerned with the characteristics of the actual journey; variables concerned with the nature of the trip; and variables concerned with the socioeconomic characteristics of the traveler.

Time Variables[1]

Discussion of the time variable in modal-choice models takes three forms; the first is concerned with which part of the journey the time variable refers to; the second is concerned with the method used to express the relationship between the times by each mode; the third deals with combinations of variables.

Total vs. Excess Journey Time. Opinion about this aspect of the form of the time variable ranges between two limiting positions, the first of which utilizes the total journey time, while the second uses the component parts of total journey time. Stopher, Lisco, and Lave all use total journey time, but Quarmby considers it more appropriate to separate in-vehicle time from time spent walking and waiting. It seems reasonable to treat time spent on different activities as different, for time spent in a car seems different from time spent waiting in a line or walking between vehicles. It is possible, however, that the intuitively perceived differences in different types of time reflect the relative inconvenience or frustration associated with the activities carried out in the time; for example, walking is more arduous than sitting, and standing in a line is more frustrating than actually traveling. In so far as the separation of journey time into its component parts allows additional variables to be included, and hence additional coefficients and values of time to be estimated, it should be made clear that differences in these values of time reflect differences in the levels of inconvenience of frustration associated with the activities (standing, walking, traveling) undertaken in the time.

As far as this study is concerned, it is desirable to distinguish the different types of time. Two categories are established: in-vehicle time and walk/wait time. Practical considerations arising from the data make it impossible to subdivide walk/wait time; this is explained by the rationalization that where subjects failed to report details of walking and waiting time, it was because they failed to perceive them in sufficient detail.

Time Differences vs. Time Ratios. The second problem concerned with the expression of the journey time variable arises from the necessity of representing that the subject considers relative journey times. Basically it is possible to express this variable as a difference or as a ratio: i.e., time by mode A — time by mode B or $\dfrac{\text{time by mode A}}{\text{time by mode B}}$. Warner has used ratios, whereas Quarmby, Lisco, Lave, and Stopher have used differences. Since the model is an attempt to represent actual behavior, it seems better to use differences, for the traveler is more likely to perceive relative times in terms of differences (faster or slower) than in terms of ratios. A preference for a difference formulation is based, therefore, upon a subjective judgment about the way in which people think.

It may be concluded, then, that the journey time variables can be expressed in a number of ways. It is felt that the most useful formulations will be those that express the relativity of the times in terms of differences and which treat the component parts of the journey as separate entitites.

Variable Combinations. The question of how to combine journey time variables arises only when journey time is divided into walking, waiting, and in-vehicle time. The problem consists of deciding in which way the traveler perceives the characteristics of his journey. The limiting cases are those in which he perceives either the total journey time difference alone or the differences in each part of the journey, i.e., walking, waiting, and in-vehicle. Clearly, other possibilities make up the intermediate positions. For example, if a traveler were not influenced by waiting time, but reacted strongly to walking time, he might consider only the differences in total journey time and in walking time, ignoring any difference in waiting time. In the absence of information on travelers' preferences, the choice of variables is best carried out experimentally. In this study the problem is simplified by the fact that walking and waiting time are not separate variables.

Relative Differences. It is argued above that the difference formulation is the most appropriate. However, the fact that this study is concerned with explaining modal choices for a medium-range intercity trip raises the possibility of a new formulation of the time variable which comprises some aspects of the ratio formulation. The trips under consideration are longer than those examined in previous studies, with the result that two major differences occur: the relationship between the time difference and the trip lengths is different; and the range of trip lengths is different.

These differences introduce the possibility that the time variable should reflect the relationship between the time difference and the trip length. The introduction of this relationship into the time variable reflects the postulate that the traveler, in selecting a mode, is influenced by more than the absolute time difference. For example, if a traveler can save twenty minutes by choosing a faster mode, the time-saving may act as a greater stimulus if the total journey time is 60 minutes than if it is 120 minutes; in other words, five minutes saved on a ten-minute journey may be important, whereas it is unlikely to be important on a four-hour journey.

Figure 4-1 shows an example. Because it is difficult to say whether a traveler would base his assessment on the faster or slower time, it has been decided to use the mean of the two times to indicate total journey time. If the time difference (ΔT) alone was considered, the two situations would be identical, but the use of the difference ratio produces different results. In Situation 1 it gives a higher value than in Situation 2, reflecting the postulate that a given time difference will be valued more highly, the shorter the journey. It is felt that the length of the intercity trip may make this an important consideration.

	Situation 1	Situation 2
TA	60	135
TB	90	165
ΔT	30	30
$\dfrac{\Delta T}{\frac{TA + TB}{2}}$	0.40	0.20

Figure 4-1. Relative Difference Formulation

Cost Variables

Much of the above discussion of the form of the journey time variable is applicable to the formulation of the journey cost variable. It is possible to consider differences or ratios, or difference ratios; it is also possible to treat only the "whole journey" cost as a variable or to subdivide it and work with various combinations of its component parts. As in the case of the time variable, it is impossible to provide a sound objective justification for the selection of any formulation of the cost variable; a subjective judgment is necessary. As in the case of the time variable, the difference formulation seems to reflect the postulated behavior of the traveler better than the ratio formulation; it is possible, however, that the difference-ratio formulation may be even better.

Aggregate vs. Per Capita Cost. A further problem in the consideration of the cost variable arises from the way in which costs are reported. The cost of traveling by car is reported in terms of the cost of driving from origin to destination; on the other hand, the cost of traveling by train is reported in terms of the individual fares; thus, a discrepancy arises in the data. In theory, this problem may be resolved in two ways: either by converting the car cost into a per capita cost or by converting the train cost into a group (or party) cost. The choice of solution depends on whether the unit of analysis is the individual (in which case the former solution is appropriate) or the traveling party (in which case the latter solution should be adopted). It is not clear whether the behavior of the traveler is better represented by treating him as an individual or as the decision-maker for a party. It is possible that the mode-choice decisions are influenced by the cost of travel for the party, but in this study it is unclear whether reported costs for car travelers refer to the individual or the party. Since it is clear that costs by train are reported on an individual basis, it is assumed that the car costs are also, and all costs are treated as individual costs.

Other Trip Variables

Other variables that might be considered in this section relate to the comfort, safety, or reliability of travel, which might be thought of as contributing to the generalized cost of travel. Measures of such mode characteristics are difficult to obtain and are usually based on psychological scaling techniques: the subject is asked to indicate on a scale the relative level of comfort, safety, and so forth. A further problem arises from the subjective nature of the measures attempted: although allowances may be made for the influences of the variables on choice, there is no way to predict how future travelers will scale comfort, safety, etc..

For this study a variable has been derived that attempts to take into account the less tangible attributes of each mode. It was felt that, in a model of choice behavior, the choice procedure might be better represented by a variable that indicated the convenience of travel by each mode, rather than one representing a subjective evaluation of the level of comfort or safety. With this aim in mind, a measure was sought that would be a proxy for the amount of difficulty associated with making a given journey by each mode. The method of representing such a variable involves assigning a unit to each section of the journey; the sum of the journey units represents the difficulty of undertaking the journey. Figure 4-2 shows an example of the calculation of this variable.

Car Journey		Train Journey	
Walk to car	1 unit	Walk to bus	1 unit
Drive	1 unit	Wait for bus	1 unit
Park	1 unit	Travel on bus	1 unit
Walk to destination	1 unit	Walk to train	1 unit
		Wait for train	1 unit
		Travel on train	1 unit
		Walk to destination	1 unit
Totals	4 units		7 units

Figure 4-2. Journey Unit Variable

This figure demonstrates the basic mechanism by which this variable is calculated. The most interesting feature of this variable is its flexibility. Should walking be thought more "inconvenient" than waiting, the units assigned to walking could be changed. A French study[2] shows that walking time is treated as twice travel time and waiting time as three times: on this basis the traveling sections of the journey could be assigned one unit, the walking sections two units, and the waiting sections three units. The variable is flexible enough to admit a number of different assumptions about the relative levels of "inconvenience" associated with traveling in a vehicle, walking, and waiting.

It is also possible to incorporate in the variable other features of inconvenience. It will be generally agreed that children add to the inconvenience of travel, although the degree may depend upon the mode. One possibility would be to multiply the units assigned to each part of the journey by the number of children in the party (or by the number of children per adult in the party). It may be thought that as the number of children increases, the relationship should be more complex than a simple multiplication, perhaps becoming an exponential function in extreme cases.

Another factor that adds to the inconvenience of a journey is luggage; walking sections of the journey are particularly susceptible to this increase in inconvenience. The flexibility of this variable would allow adjustments for the number of pieces of luggage carried.

A combination of luggage and children could be built into the model, although the precise form of the adjustment would be open to discussion. At present, not enough is known about the opinions of travelers on the relative effects of children, luggage, etc., on the inconvenience of a journey for such detailed adjustments to be made. Moreover, in this study, data was not collected on the amount of luggage carried. The variable will, therefore, be used in its simple form. Following the discussion on the time and cost variables, it will be expressed as a difference.

Journey Purpose

Discussion of the journey-purpose variable introduces the series of variables that describe the nature rather than the characteristics of the journey. One of the principal aims of this study is to extend knowledge on modal choices and values of time beyond those derived from commuting trips. Not only may values of time be different for different trip purposes, but also the factors influencing choice of mode may either be different or operate in a different way. It is unlikely that a model will be derived to explain modal choices for all journey purposes; therefore, it is proposed to subclassify the data into categories representing the different journey purposes and to analyze the social-recreational travel subset.

Journey Frequency

The effect of the frequency with which a journey is made on the modal-choice process is unknown. Intuitively it is felt that this variable may operate indirectly by way of a type of learning process. In other words, the traveler may make a decision on the basis of information available to him, but he may modify that decision in the light of further information that can only become available as a result of regular travel. In order to investigate the effect of this variable, it is necessary to subclassify the data by frequency, but the data will not support such a subclassification. Such an investigation is recommended as an area for future research.

Mode-Pairs

The only reasonable method of dealing with the different mode-pairs arising from a tri-modal sample is to treat them separately. It is not feasible to deal with car-train and car-bus choices in the same analysis. Extensions of the statistical techniques for handling binary choice situations have been developed and may be used in future studies, but the data collection for this study restricted itself to information on two modes only; it is, therefore, not possible to utilize the multinomial extensions of binary choice analysis techniques. As the automobile-train choice is the most important, it will be the subject of analysis.

Socioeconomic Variables

Previous studies have incorporated a number of variables that reflect the characteristics of the subject or his household. A discussion of the more important ones follows.

Age/Sex. The age and sex of the subject have been included in a number of models, on the grounds that they improve the fit of the model. Lave concludes:

It is possible that there is a systematic relationship between the shape of a commuter's preference function and his age and sex. But there is no easy way of predicting their effect and it is difficult to relate their . . . coefficients to any specific real-world interpretation. Accordingly we can only rely on fitting considerations as a guide for the inclusion or exclusion of these variables.[3]

This statement adequately sums up the position: age and sex will, therefore, be included only if their inclusion improves the fit of the model.

Competition for Use of Car. Several attempts have been made to construct a variable that would reflect the demand for the car within the household. Such variables are usually derived from some combination of the number of drivers and the number of cars in the household. Unfortunately, in this study the section of the questionnaire which asked for this information was badly worded, with the result that it is uncertain whether the respondents included themselves and their car in the total number of drivers and cars in the household. The information collected is, therefore, unreliable, and the best approximation to a demand for the car variable which can be salvaged is the "number of adults" in the household.

The loss of this variable is not crucial, since the variables used in the past have failed to mirror fairly the nature of the intrafamily demand for the use of the family's car stock. The bargaining procedures are undoubtedly considerably more complex than can be represented by the simple ratio of cars to drivers.

Car Ownership. The problem of a car-ownership variable does not arise in this study as all respondents in the final data set report a meaningful choice between car and train. Effectively, those train travelers who do not own a car reported "don't know" when asked for details of the journey by car and were thus eliminated from the sample. Presumably, such respondents did not perceive the car as a real choice and should not, therefore, be included in a car-train choice subgroup. It is possible to argue that their choice alternatives included buying a car and that they should be included, with the appropriate adjustment to a car costs to take account of the average as opposed to the marginal costs. Nevertheless, it is felt that the proper procedure should not involve attributing to respondents a choice that they do not themselves perceive.

Size of Traveling Party. A variable that has not been used in the commuter studies, but which may be meaningful in the context of a leisure trip, particularly over a longer distance, is the size of the traveling party. It can be postulated that as the size of the party increases, different modes have advantages, and therefore, variables should be included which reflect the size of the traveling party. The data-collection exercise obtained data on the number of adults and the number of children in each party, and these variables will be included. In addition, it is suggested that it is the number of children under the control of each adult which is of importance, and a third variable will be constructed (number of children/number of adults) to take account of this possibility.

Income. It is generally agreed that the level of a subject's income will affect his choice of travel mode; it is further agreed that this effect is most difficult to determine. In previous studies the income variable has been handled in three main ways.

By substratification. Many analysts believe that the effect of the income variable is essentially different from that of the trip characteristic variables; rather than influencing the choice made, it fundamentally affects the choice process. The implications of this standpoint are that each income group has a decision process that must be modeled separately. It is possible that the same variables with different coefficients will explain choices for all income groups; it is equally possible that the variables may be substantially different from group to group.

The operational result of this point of view is that the sample is divided into income groups that are analyzed as distinct samples. An advantage of this method is that the relationship between the coefficients and income can be readily assessed. A strong disadvantage is that a large sample is required to make such stratifications possible; this is a problem in a commuter study, but in a multipurpose study the problems are exacerbated, as it is necessary to carry out two stratifications, first by journey purpose and then by income. It was the intention of this study to collect a sample of sufficient size to allow the two stratifications.

As a variable. Although some studies have included income as a variable,[4] it is difficult to justify the inclusion of income as a direct variable, except on the grounds that it improves the fit to the model. The use of a crude-income variable does not do justice to the complex manner in which income affects mode choices. More will be said about this process at a later point. At the moment, suffice it to say that the use of the crude income variable is not advocated.

In combination with other variables. Attempts to explain the complexities of the process by which income affects modal choices have led some analysts[5] to conclude that income operates through or in conjunction with other variables, such as cost, time, and comfort. In some cases, it is argued, the cost difference is important only in relation to income, so that a given cost difference will produce different reactions in a higher-income-group traveler than in one from a lower-income group; a suggested solution is to combine the income and cost variables to produce a new variable, say, the ratio of cost difference to income. In other cases, it is argued that it is the time difference that is perceived differently by different income groups, so that a given time difference will affect the higher-income-group traveler more because he values his time more highly (assuming that the value of time rises with income). In yet other instances, it is claimed that the higher-income-group traveler is more susceptible to differences in comfort, and so a new variable is derived, composed of the comfort variable multiplied by income.

With the exception of stratification, these methods of dealing with income are inadequate, for they fail to take account of the true complexity of the effect of income on the choice process. Clearly, a great deal of work would be necessary to solve this problem. Nevertheless, an attempt will be made to explain the process and to formulate the income variable in the way that best reflects its effect.

Two basic patterns appear tenable. One holds that the influence of income on the choice process is so fundamental that it is necessary to deal with each income group separately. Insofar as the data permits, this approach was utilized. The other position, enforced perhaps by the restrictions of most samples, attempts to develop a variable that accurately reflects the influence of income. Most attempts to do this, however, are conceptually unsatisfactory, because they consider the effect of income on individual variables, rather than on the choice process as a whole. The basis of the modal-choice modeling process is a trade-off situation, be it between time and money, comfort and money, or any other combinations of variables. In a trade-off situation the choice of one alternative involves giving up one commodity for another; a time-money trade-off would involve two alternatives, one of which was faster, but more expensive, than the other, so that a choice would necessitate either giving up money to save time, or giving up time to save money. Any group of travelers faced with such a choice can be divided into time-choosers, who spend money to save time, and money-choosers, who expend time to save money. The comfort-money trade-off operates in exactly the same way: some people will spend money for extra comfort, while others will suffer discomfort to save money.

What, then, is the effect of income on this choice process? It is argued that the effect of a higher income is to lessen the probability that the person will be a money-chooser. In other words, the higher the income group to which a subject belongs, the more likely he is to choose the alternative that involves spending money to save time and increase comfort.

This contention embodies the arguments raised above; a traveler from a higher-income group will be less influenced by a given cost difference (because he has money available); he will be more influenced by a given time difference (because, it is assumed, his time is more valuable); and he will be more influenced by a difference in comfort (because he is more susceptible to comfort). This formulation can be simply stated: The higher the income group to which a subject belongs, the less likely he is to be a money-chooser in any given trade-off situation.

The form of the relationship between the probability of being a money-chooser and income is open to discussion, but it seems possible that it may take the form of the sigmoid curve discussed earlier in this thesis. The relationship is shown in Figure 4-3; $P(x)$ is the probability of being a money-chooser.

This relationship means that those in the highest income group have a 0.0 probability of being money-choosers; those in the lowest income group have a 1.0 probability of being money-choosers. It is, of course, possible to postulate a threshold effect in the relationship, such as the one described by Lisco, by

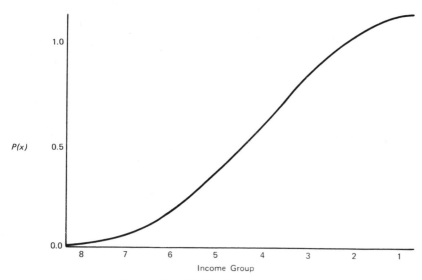

Figure 4-3. Choice-Income Relationship

setting the 1.0 probability for $P(x)$ at, say, the third income level, and equally to set the 0.0 probability for $P(x)$ at the seventh income level (Figure 4-4).

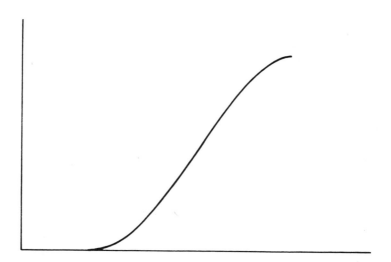

Figure 4-4. Choice-Income Relationship with Budget Constraint

Such a relationship would reflect the fact that budgetary constraints would compel those in the lowest income groups (1, 2, and 3) to be money-choosers, whereas financial affluence would give groups 7 and 8 the freedom never to be money-choosers. The form of this relationship could be tested empirically, but such testing is beyond the scope of this study.

To consider the relationship between income and the probability of being a money-chooser is, however, to consider only a part of the problem; the whole problem is ascertaining the relationship between income and the probability of choosing a given mode. If it can be assumed that in a given situation one mode is always cheaper than the other, then to be a money-chooser determines which mode will be chosen over the whole sample. This assumption is probably justified in a study of commuting trips, where the trips are relatively homogeneous and it can be claimed with some certainty that public transport is cheaper (although slower) than car. In such a case, to be a money-chooser is to be a public-transport-chooser and relationship:

$$P \text{ (money-chooser)} = F \text{ (income)}$$

can be translated directly into:

$$P \text{ (public-transport-chooser)} = F \text{ (income)}$$

and the functional relationships will remain unchanged.

In a study of intercity trips or any other type of nonhomogeneous trips, the picture is less clear. The nonhomogeneity may mean that, in some cases, the car is both faster and cheaper than public transport. Consider a suburb to suburb trip, as pictured in Figure 4-5.

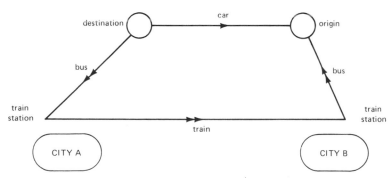

Figure 4-5. Homogeneity of Journeys

For a journey such as the one presented schematically above, it is possible that the car-cost would be less than the public-transport cost. It would be necessary first to estimate the relationship between the choice of money and income and then to examine the data to determine whether a money-chooser, for example, can be equated with a public-transport-chooser.

Such an examination is, however, beyond the scope of this study, which will be limited to dealing with the income variable by stratification.

The discussion of the income variable concludes the review of the variables that could be included in the modal-choice model and the forms they could take. It is possible to collect other variables; for example, Thomas collected information on attitudes to roads and on physical aspects of a route. Since it is always possible to think of new variables, the list under review is not exhaustive. It covers, however, the important variables and the main points of discussion about their possible forms. It is necessary to remember that the collection of data is limited by factors other than the inability to invent new variables. The necessity of obtaining voluntary cooperation is one such factor; the quality of data required is another. It is also advisable to be able to explain the nature of the influence of a variable on the choice being considered; although an increase in the number of explanatory variables will often increase the explanatory power of a model, addition of variables for this reason is a bad analytical procedure.

The use of the variables discussed above has been justified in terms of observable behavior, and their use in a modal-choice model should explain choice behavior, if it can be explained in terms of the basic postulates of the model.

Combinations of Variables

Given the magnitude of the task of computing a model utilizing all the possible combinations of variables, it is necessary to set out a procedure for combining the variables. This procedure, of course, can be no more than a guideline, since the inclusion and exclusion of variables depends to a large extent on the judgment of the analyst and the interim results. Two main factors influence the development of the model content. First, the aim of this study is to derive a value of time from the modal-choice model; this means that the time and cost variables should, if possible, be included in the model. Second, *ceteris paribus*, a simpler model is better than a complex one, where "simpler" means containing fewer variables. The latter factor leads to the procedure of building up a model by the addition of new variables. Such a procedure is preferable to starting with a large number of variables and eliminating insignificant ones; the former factor leads to the choice of the time and cost variables as the initial combination.

5

Data Collection

This chapter explains the method used to collect the data used to develop and test the model outlined in the preceding chapters. A number of tasks are involved in the collection of data, and they tend to interact in a complex fashion. This chapter is based on the outline of survey development procedure presented in Figure 5-1. This figure describes the basic procedure for developing a survey, and each task will be considered in turn; clearly, decisions made about the way in which one task is to be performed will affect and effectively place restraints upon other tasks. For example, the choice of a sample is likely to restrain the range of data-collection methods that can be used.

It should be made clear at the outset that the type of data required for this study had to be collected by survey methods. Such data is collected neither by government agencies nor by consultants carrying out transport studies. As the study involves investigating the behavior of individuals, it was necessary to approach the subjects concerned by means of a questionnaire. How the subjects were chosen, how they were approached, and what questions they were asked will now be discussed.

Definition of Sample

The preliminary notion of the sample stems from the fact that the study is concerned with explaining the modal-choice behavior of intercity travelers. The sample, therefore, had to be composed of people traveling between two cities. The choice of travel route was constrained by certain criteria that an appropriate route should meet:

1. The route should link two sizable centers of population.

2. It should offer adequate travel facilities between the centers of population in order to ensure a meaningful choice of travel mode.

3. The methods of transport should be amenable to survey examination without undue difficulty or cost.

Location of personnel and a knowledge of the facilities available led to the choice of Edinburgh as one end of the travel route. London was considered as the other, but was rejected as the identification and surveying of road travelers appeared to pose problems that could only be overcome by the application of much time and money. In short, road traffic to London could leave Edinburgh by one of three routes, and the number of London-bound travelers in the normal

69

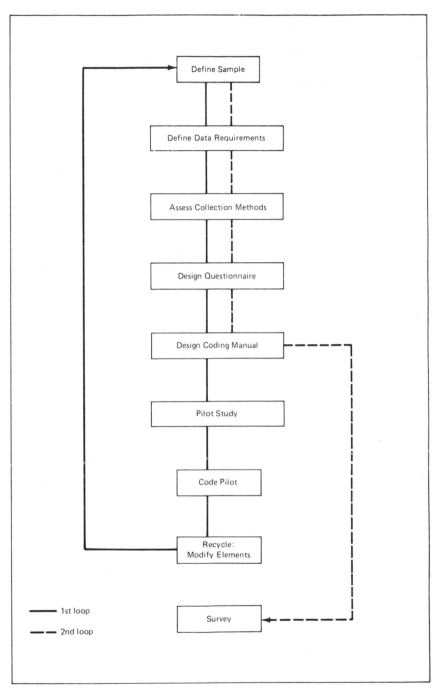

Figure 5-1. Survey Design Tasks

traffic flow was likely to be so small as to necessitate prolonged surveying to collect a sample of reasonable size.

The Edinburgh-Glasgow route, on the other hand, had the advantage that the two cities are linked by one main road, so that the physical problems of surveying would be reduced considerably. Moreover, the shorter distance meant that it would be likely that a larger proportion of the total traffic would be making a trip of the type with which the study was concerned. The shorter distance also meant that the range of journey purposes was more interesting, including a higher proportion of recreational and commuting trips. From the administrative point of view, the restriction of the study to Scottish roads and railways meant that the number of public agencies involved was reduced.

The next problem was to define what was meant by an Edinburgh-Glasgow trip. Since the cities are each to some extent the centers of a conurbation (towns adjoin them which are distinct municipal units) and the center of population for a larger area, it seemed unreasonable to restrict the sample to travelers from a point within the boundaries of one town to another point within the boundaries of the other. Since boundaries are sometimes derived arbitrarily, it would be possible for a trip from just outside one boundary to be more relevant to the study than one from just inside the boundary. If the sample area was to be defined in terms of the "catchment areas" of the two cities, the problem of defining the "catchment area" arose. The problem was aggravated by the question of the homogeneity of journeys, which is expressed pictorially in Figure 5-2.

Is the journey from A to B by car (solid line) which passes through the one city homogeneous with the journey by rail between A and B (dashed line) which passes through both cities? As a person traveling from A to B and choosing between road and rail would have to make a choice between the two journeys, it was decided that they could be considered homogeneous in the sense that to the subject they represented alternative methods of making the same journey.

This decision solves the problem of setting outer limits to the catchment area. Since the study was concerned with travel between the two cities, a relevant journey was defined as one where the public transport alternative involved traveling into the city itself to utilize the intercity rail link. Therefore, all rail travelers were automatically making relevant journeys, with the obvious exception of those descending at intermediate railway stations. The remaining problem was to allocate the car travelers; they were allocated to the sample if their public transport alternative journey involved using the intercity rail link. Thus, a car traveler making a journey from Edinburgh to a destination near Glasgow served by an intermediate train station would be excluded from the sample.

The derivation of the catchment area boundary is shown in Figure 5-3. Point A is within the catchment area, because the rail traveler to Edinburgh would make the journey into the city center to catch the train; point B is outside the

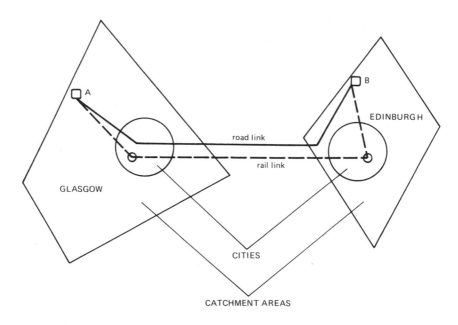

Figure 5-2. Trip Homogeneity Definition

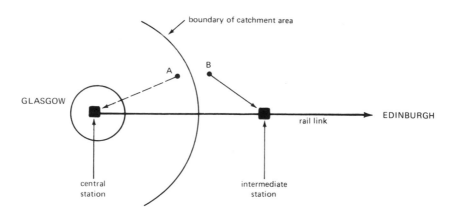

Figure 5-3. Catchment Area Boundary Definition

73

catchment area, because the rail traveler to Edinburgh would go to an intermediate station to catch the train. In theory, there would be a point between A and B where the traveler would be indifferent between stations; in practice, the subject reports a journey to one or the other station. Thus, the exact location of the boundary line raises no operational problems.

Definition of Data Requirements

The data requirements for this study involved two decisions: the data to be collected from each respondent; and the total number of responses to be collected.

Choice of Variables

The choice of the variables on which data was to be collected was strongly influenced by discussions of the type set out in the previous chapter, although it should be made clear that the preceding discussion represents the crystallization of much work on the topic. At the planning stage of the study, it was less clear which variables would finally be included in the model, and, therefore, data was collected on some variables that were subsequently abandoned. Moreover, some information was used at the coding stage to specify and modify journeys made and does not appear in the model as an explanatory variable.

The data collected was classified into four categories:

Actual trip information, or information on the origin, destination, times, costs, and other features of the trip being undertaken.

Alternative trip information, or information on the same features of the trip as it would have been made by the second-best mode.

Features of the trip, or information on the purpose and frequency of the trip.

Socioeconomic information, or information on the socioeconomic characteristics of the subject.

Choice of Sample Size

In this study, the choice of sample size was determined partly by the nature of the sample and partly by the aims of the study. In the first place, the sample was to be made up of people traveling for different purposes, with the result of an obligatory stratification of the sample by trip purpose. Furthermore, one of the aims of the study is to examine the way in which different income groups make modal choices and value time. The achievement of this goal involves a further stratification by income group. Thus, the sample collected had to be large enough to allow these stratifications to be made and still leave enough respondents in each cell to make the statistical analysis a viable proposition.

Choice of Collection Method

Having decided on the information to be solicited from respondents and identified the respondents, the next problem was actually obtaining the information. Given the use of a questionnaire, two basic approaches seemed possible. The travelers could be interviewed in their homes or in transit. The former method would involve selecting a sample of households and questioning the occupants on the journeys made during a specified time period. This method, while suitable for an investigation of commuting trips, was unsuitable for investigating an intercity trip. Most people make a daily journey to work; fewer people make intercity trips, with the result that a large sample of households might be expected to yield only a small sample of people who had made a specific intercity trip. The collection of a sample of reasonable size would be a time-consuming and expensive operation. (It should be mentioned, however, that it is relatively easy to interview people in their homes.)

To ensure that the people being interviewed did, in fact, fit the sample, it was better to question people as they traveled. This method, however, while ideal in terms of identifying the sample, is fraught with difficulties which arise from the fact that travelers are likely to be in a hurry, laden with luggage and/or children, and, therefore unresponsive to an interviewer. It was felt, however, that those problems would be overcome, whereas those of identifying a sample from a random sample of households could not. Thus, it was decided to approach travelers in transit.

Having made this decision, the problem was redefined as that of discovering the best method of obtaining information from travelers by bus, train, and car. Four methods of administering questionnaires were considered as most likely to produce a favorable response; each has advantages and disadvantages.

Method 1. The use of a trained interviewer to question subjects has the advantage associated with depth interviewing, a low probability of the subject giving incorrect or inadequate answers due to a misunderstanding of the questions. This method has disadvantages when applied to a travel situation. On a bus or train, subjects may dislike being questioned in front of other travelers and may refuse to cooperate or provide false information. Interviewing car travelers in this way would entail detaining them at the roadside for a considerable time, and it seems unlikely that drivers thus antagonized would provide good information.

Method 2. Obtaining the interest and cooperation of a subject in transit and obtaining the required information in a follow-up interview in the privacy of his home avoids the disadvantages of in-transit interviewing, but raises two others. First, it is impossible to preserve the anonymity of the respondent, since it is necessary to obtain his name and address. The knowledge that he can be readily identified with his responses may deter the subject from cooperating. Assurances that the subject-response link will be destroyed after the interview are unlikely to be satisfactory. The second disadvantage is that this method involves all the expense of a survey to identify the subjects plus the additional expense of the home interviews.

Method 3. The on-mode distribution and collection of questionnaires was the third method under consideration. The questionnaires are self-administered by the subject as he travels and are collected before he leaves the train or bus. (Obviously, this method is not applicable to car travelers.) This method has the advantages that the respondent can fill in his questionnaire in privacy without fear of his replies being overheard; moreover, the time delays and the expense associated with follow-up interviews are avoided. The disadvantage is that the questionnaires are self-administered, with the risk that the subjects may misunderstand the questions and provide useless information; this disadvantage is mitigated if a distributor is available to help respondents should difficulties arise.

Method 4. Distributing questionnaires for postal return in prepaid reply envelopes was the final method. It is thought unlikely that train or bus surveys would require such a method, but it should be useful for car surveys. The disadvantages of this method are those associated with self-administered questionnaires plus the uncertainty of the response. The advantage is speed of distribution.

A further method by which interviewers would question people at the bus and train stations was rejected on the grounds that people looking for and/or hurrying to trains or buses are unlikely to be responsive to interviewers; furthermore, the distribution of arrivals would mean that even a team of interviewers could only interview a small proportion of travelers. For these reasons, only methods that approached the traveler actually on the train or bus were considered.

Since little was known about the reactions of travelers when presented with a questionnaire while in transit, Methods 1, 3, and 4 were tested in a pilot study. (Method 2 was abandoned as raising too many problems.) The results of the pilot study were most interesting.

Before the pilot study, it was thought that the use of professional interviewers to question respondents would have the advantage of obtaining fewer incorrect or inadequate answers. On the other hand, it was thought that respondents might dislike being questioned in open buses or trains, and that car drivers might be antagonized by the delay occasioned by a complete interview. In practice, it was found that the interviewer's desire to complete interviews in a hurry, and thus maximize the number of interviews achieved in a given time, led to inaccuracies. It was also noticed that some of the information reported was incomplete, perhaps owing to the fact that the desire for speed induced some interviewers to omit full explanations of some of the questions. On the other hand, interviewers reported that respondents were quite willing to answer the questions (including those on income and other personal data), even when they could be overheard by other people in adjacent seats. Moreover, car drivers proved willing to be interviewed for up to ten minutes, and the number of interviews terminated or refused owing to a lack of time on the part of the respondent were very few.

The "on-mode distribution" method was originally thought to have the advantage of allowing much greater coverage than direct interviewing, while suffering from the disadvantages of incompleteness in the interviews, and a large number of refusals due to the complexity of the questionnaire. In practice, it was found that far from avoiding the complex questions and restricting themselves to those requiring a yes/no answer, the respondents were, on the whole, quite willing to provide detailed accounts of their journeys. It was noted that the response from the train passengers tended to be more comprehensive than that from the bus passengers.

The "postal" method of distributing questionnaires with prepaid reply envelopes in which to return them was most applicable to the roadside survey of car travelers and provided excellent coverage, especially at peak traffic times when the questionnaires could be given to a high proportion of drivers with minimum delays. Most drivers seemed pleased that efforts were obviously being made to minimize delays, and it is felt that the creation of such good will between driver and survey is most important.

In conclusion, the pilot study demonstrated that direct interviewing is possible on buses, on trains, and at the roadside. The coverage to be obtained by this method is, however, somewhat limited, and it is thought undesirable to use this method when the field to be covered is large. "On-mode distribution" showed that travelers both would and could fill in a questionnaire while traveling. The "postal" method proved to be a most efficient method of getting questionnaires to the public and is most useful when it is important to minimize the delay to the respondent.

As a result of the experience of the pilot study, it was decided to use a different method for each mode of transport: the "postal" method for car travelers, the "on-mode distribution" method for the train travelers, and the "direct interviewing" method for the bus travelers.

Questionnaire Design

It is difficult to report the process by which a questionnaire is designed, since it involves a continuous cycle of discussions and modifications which converges on the final form. It is intended, therefore, to state briefly the factors that influenced the development of the questionnaire.

Three main conditions had to be met by the questionnaire:

1. It had to elicit information on a number of topics as efficiently as possible.

2. It had to be sufficiently adaptable to cover three modes of transport from which different information was sought.

3. It had to record this information in a manner which was consistent from mode to mode in order to facilitate the subsequent coding operations.

To meet the first condition it was necessary to find a way of asking the traveler a large number of questions in a manner that would make clear the information required, and link the information in such a way that the subject could see a clear pattern in the questions. To achieve these aims, the questionnaire was divided into four sections to obtain information on: the journey being undertaken; the subject's habits and attitudes; the alternative journey by the second-best mode; and the subject's socioeconomic characteristics.

The questions attempted to take the subject through his journey in a logical manner, since it was felt that this would produce better information and avoid any frustration that might be caused by jumping from one part to another. Following the journey details were questions on the nature of the journey (its purpose, frequency, etc.) and the subject's attitude to various aspects of the chosen mode. Next, it was put to the traveler that he might, for some reason, have to travel by another mode and he was questioned on the hypothetical, second-best journey. Finally, he was asked to provide information on his socioeconomic characteristics.

In order to meet condition (2), one basic questionnaire was produced with a basic set of questions common to all three modes. Alternative wordings, substitute and additional questions were then designed so that the questionnaire would be modified to suit each mode. Moreover, the use of different methods of distribution meant that the form of the questionnaire had to be modified; for example, the self-administered and the interviewer-administered questionnaires needed different instructions.

To record the information in a consistent form for coding purposes was difficult. Ideally, the coding manual should have been drawn up when the questionnaire was designed; this was not done owing to pressure of time, with the result that the coding column of the questionnaire was inadequately specified. This meant that the information had to be put into a consistent, usable form at the coding stage, which was consequently made both more tedious and more time-consuming.

The factors described above are, then, those considered when the questionnaire was designed. The development process cannot be defined precisely, being by nature a pragmatic procedure that involves putting each question to a large number of people in an attempt to find a form that minimizes misunderstandings.

The Pilot Study

The aims of the pilot study were: (1) to test methods of administering the questionnaires; (2) to obtain information about the population; (3) to test the questionnaires; and (4) to test the operational feature of the survey.

1. This has already been discussed in the section on collection methods.

2. As little was known about the population to be investigated, it was desirable in the pilot study to obtain as much information as was possible. This was particularly important since the final sample size, and the survey effort necessary to produce it, can only be determined in light of prior knowledge of the population. This is especially applicable given the stratifications it is intended to perform. If the proportion of the population in each subgroup is unknown, it is impossible to devise a sample that will result in each subgroup containing a number of observations large enough to make statistical analysis feasible.

3. While the ad hoc procedures of questionnaire design and development can produce a good draft questionnaire, it is only possible to assess a questionnaire under working conditions. It is then that what was thought to be a difficult question is answered correctly and an apparently easy question produces all manner of misunderstandings. The pilot study showed which questions were frequently misunderstood and allowed them to be modified.

4. It is not intended to describe the operational features of which the pilot study was a test, as they will be set out in a later section which describes the survey. It is sufficient to say that they were mainly of a logistic nature: requirements of staff, materials, etc., and the feasibility of work loads, timings, and so forth.

Refinements

As a result of the pilot study, it was possible to reach a decision on the sample size and necessary survey effort, on the most appropriate methods of administering the questionnaires, on the necessary modifications to the questionnaires, and on the physical operation of the survey. It cannot be stressed too strongly that the pilot study was a vital part of the process leading up to the main survey.

The Survey

The main survey was carried out during the period September 13-28, 1969, inclusive. The time was designed to avoid local holidays, while utilizing the long hours of daylight to enable the roadside interviewing to continue for a reasonable length of time each day. The roadside survey occupied the first ten days of the survey period, and the rail and bus surveys took place during the last seven days.

The Car Survey

The roadside survey was a comprehensive operation covering all traffic traveling on the A8 road in a westward direction, from Edinburgh to Glasgow.[1] It consisted of two parts: an origin-destination survey of all travelers, and a more detailed investigation of the journey and socio-economic characteristics of selected travelers.

The field-interview station was set up on the A8 road approximately eight miles west of Edinburgh (Figure 5-4). The location was selected largely on pragmatic grounds (few sites were really suitable), but it met a number of important conditions. It was located at a "bottleneck" point in that it was close to Edinburgh and yet avoided intracity traffic movements; moreover, it was far enough away to pick up traffic filtering onto the A8 by local roads from southern suburbs or small towns west of the city. The road at this point had four lanes, with a central reservation and also a long lay-by, which meant that the survey could take place with minimum delays to traffic.

The location is pictured diagramatically in Figure 5-5, which shows the overall picture, and Figure 5-6, which shows the site in more detail.

Traffic approaching the site met warning signs approximately half a mile before the site; the signs warned them to slow down and that they were approaching a survey. The two lanes of traffic were then merged into one lane in order to simplify the task of the policeman who had to separate the private vehicles from the commercial vehicles (Table 5-1). The latter were allowed to pass in the outside lane, which was also used as a safety valve at times of congestion; if the bay area became congested, all traffic was directed into the outside lane, thus allowing the congestion to clear, preventing a build-up of traffic and minimizing delays. (This procedure was only adopted during peak periods and, owing to the efficiency of the system, only infrequently.) Just before the vehicles were separated into two groups, they were counted and classified into various categories of vehicles. As a stream of cars approached the first bay, it was the task of the first traffic marshal to ascertain whether or not the bay was empty and, if it was, to direct the first car into the bay. He would then repeat the procedure for the second bay. If both his bays were full, he would pass the cars on to the second marshal, who would attempt to feed them into bays three and/or four. This system continued until the cars reached the policeman controlling bays seven and eight. It was found advisable to have the policeman at the end of the line sweep up the odd car who refused to obey the signals of the traffic marshals. Thus, it was the responsibility of each marshal to keep his bays full whenever possible; it was also sometimes necessary for him to assist cars to leave his bays into a stream of traffic.

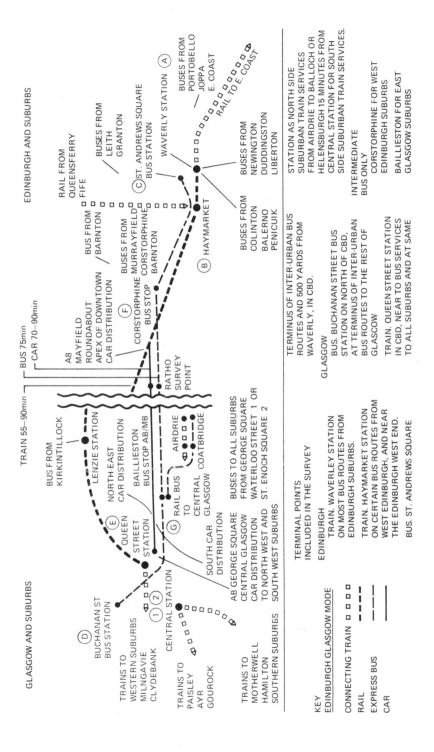

Figure 5-4. Site Location

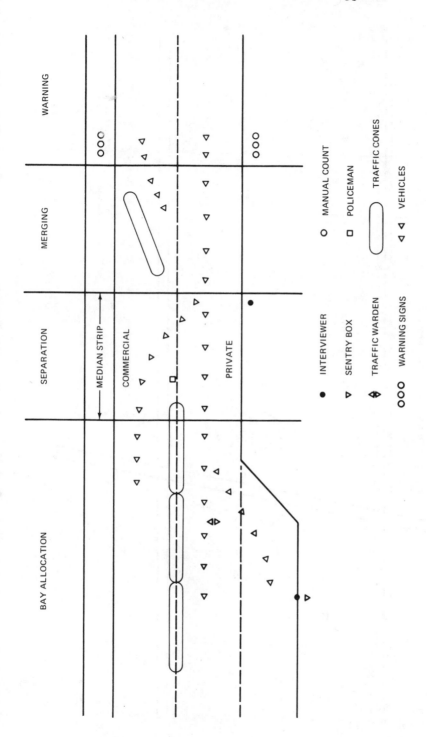

Figure 5-5. Roadside Survey—Overall View

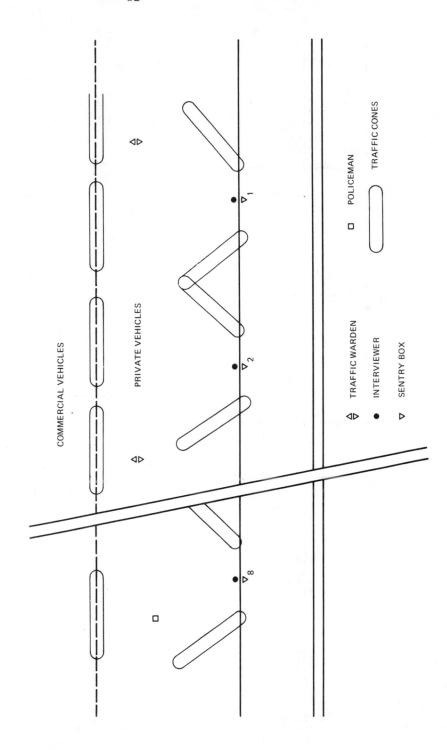

Figure 5-6. Roadside Survey—Interview Site

Table 5-1
Vehicle Classification

Vehicles to be stopped:
a) All ordinary cars, with or without trailers
b) All station wagons and vans used for private purposes
c) All three-wheeled cars
d) Invalid carriages
e) Taxis
f) Motorcycles, motorcycle combinations, mopeds, scooters

Vehicles not to be stopped
a) All commercial vans and trucks
b) All lorries
c) Buses and coaches (private or scheduled)
d) Pedal cycles
e) Emergency vehicles (fire engines, ambulances, police cars)
f) Road Rollers
g) Hand Barrows
h) Red G.P.O. Vans
i) Other unusual vehicles (e.g. tanks, or other military vehicles)
j) Horses
k) Horse-drawn carts

When a car had been directed into an interviewing bay, the interviewer approached it from the passenger, or near, side. This procedure, although contrary to established practice, was adopted for the safety of the interviewers, since it was felt that the dangers of interviewing at the driver's side, with a stream of traffic behind the interviewer, were too great. During the pilot study, a tendency to terminate an interview and step backwards into the stream of traffic was observed, and it was felt that the slight inconvenience to the motorist of having to open the passenger window or door was worth incurring in order to avoid risks to the interviewers.

The actual interview proceeded in the following manner:

1. The study was introduced and its purpose briefly explained.

2. The driver was asked for the address of his origin and destination, in as much detail as possible. The detail was necessary for coding purposes, as the street number might determine the zone to which the address was allocated. This was explained if necessary.

3. The main purpose of the journey was ascertained and allocated to one of the seven categories shown in Table 5-2.

4. The number of people traveling in the car was recorded.

This format was followed in the case of all cars stopped. The remaining part of the interview could take two forms. If both the origin and destination addresses fell within the previously defined catchment areas of the two cities, the subject was given a copy of the questionnaire which he was requested to fill in and return in the prepaid reply envelope supplied. The interview was then

Table 5-2
Trip Purposes

1. Social/Recreational:	Includes trips to private houses to visit friends or relatives; trips to restaurants, cafes, bars; trips to recreational facilities such as movies, dance-halls, football matches, public meetings; and pleasure motoring.
2. Employers Business:	Includes trips by employed and self-employed persons in the normal performance of their work, e.g. trips by doctors, salesmen, plumbers, delivery men, etc., as well as "business trips."
3. Journey To/From Work:	Persons going to or returning from their place of work, including trips home for lunch.
4. Personal Business:	Includes all trips to fulfill personal business transactions not connected with a person's employment, e.g. trips to pay bills, renew licenses, visit lawyers or doctors, etc.
5. Shopping:	Trips to buy goods for personal use.
6. Educational:	Trips to school or other places giving education at any level, including adult education. (N.B. teachers going to school to teach are on journeys to work.)
7. Other:	Any trip not covered by categories 1-6.

terminated. In cases where either the origin address or the destination address were outside the catchment areas, the interview was terminated at this point.

The roadside survey was conducted continuously for the ten-day period. Some difficulty was experienced due to bad light in the evenings, and the closing time for interviewing varied from 7:00 P.M. to 8:00 P.M. depending on the extent to which weather conditions affected the onset of dusk. Apart from one day when fog prevented any interviewing before 11:00 A.M., the survey ran from 7:00 A.M. to 8:00 P.M. daily, whether or not interviewing was practicable.

The Train Survey

The train survey was altogether a much simpler operation to mount than the car survey, as it did not require the special facilities necessitated by a roadside interviewing station. Nor did it require such a drastic disruption to the public, since the subjects were being approached as they traveled in their usual manner. The survey involved the use of self-administered questionnaires, which were distributed to all passengers, avoiding where possible passengers traveling to intermediate destinations. In practice, this was not always possible. As discussed above, it was thought undesirable to approach passengers as they were in the process of boarding the train; a count was maintained, however, at the ticket barrier, and the destination of each passenger was noted. (This was made possible by the cooperation of the ticket collectors.)

When the train pulled out of the station the distribution began. After a brief explanatory preamble, questionnaires were given to each passenger, and they were requested to fill them in.

The distributors were on hand during the journey to help people who found the questionnaire difficult and to provide encouragement (or gentle coercion) to those who seemed slow to complete the questionnaire. This was a matter for the judgment of the distributor. Cooperation was entirely voluntary and any refusal to answer, resistance, or antagonism was met by a polite but prompt termination of the interview.

The choice of trains to be sampled was such that a timetable was drawn up of every third train leaving each station. The timing of the first train surveyed was staggered from day to day so that each weekday service was covered twice during the survey period, with the exception of one or two morning and evening peak-hour trains and some "football specials," which were only covered once.

The Bus Survey

As the number of passengers traveling between the Edinburgh and Glasgow areas by bus was many fewer than by train or car, professional interviewers were used to interview passengers on the bus survey, which took place concurrently with the train survey, thus allowing the same supervisory staff and office facilities to be used for both. The questionnaire used by the professional interviewers was modified to allow information to be recorded more easily by the interviewers, visual attractiveness being unnecessary, since the passengers would not have to see the questionnaire.

In general, it was possible for a team of three interviewers to interview all passengers on the off-peak and evening buses, and eighteen to twenty people on the more fully loaded buses. A count was kept of the number of passengers traveling by the interviewing team. No count was kept on buses not being surveyed.

Interviewers worked on every second bus, beginning alternatively with the first and second morning buses, thus ensuring good coverage of all buses.

Conclusion

This chapter has explained in some detail the way in which the data collection for this study was developed and carried out. It is important to do this, since the collection of data from intercity travelers was one of the novel features of this study, and methods of presenting a complex questionnaire to a subject actually on a journey had to be developed. It must be admitted that given a greater amount of time to set up the survey, and given greater experience of survey work, certain aspects of the survey could have been improved. On the whole, considering the extent of the operation and the newness of some of the problems, it is thought that the survey work was extremely successful.

6   The Data

The use of data collected by methods such as those described in the preceding chapter involves many problems not present when "prepared" data, such as government statistics, are used. It is, therefore, incumbent upon the analyst to devote more care to the analysis of the data and to instigate data checks before proceeding to any statistical analysis. First, the crude traffic figures are presented and examined for any obvious peculiarities. Second, the derivation of the data set finally used for the statistical analysis is explained and, third, the final data set is checked against the total sample collected for any biases that may have arisen in its selection. Fourth, the final data set is checked internally by an examination of the frequency distributions of the more important variables, and by consideration of the correlation matrix. Finally, it was thought advisable to check that the data set is truly representative of an interurban trip.

Given the binary nature of the model under consideration, it was decided only to analyze (in the first instance) the choice between road and rail. This chapter, therefore, confines itself to an examination of the data collected from travelers by car and train.

The Population

The results of the counts carried out on the number of passengers traveling by all trains and on the total traffic flow during the hours of the roadside survey form the population from which the sample was selected.

The total picture of the population traveling by rail is summarized in Table 6-1. Two points are noteworthy. First the "% of trains surveyed" column shows that (excluding the less frequent Sunday service) the trains chosen to be surveyed carried between 25 percent and 37 percent of Edinburgh-Glasgow passengers in the westbound direction and between 28 percent and 34 percent of Glasgow-Edinburgh passengers in the eastbound direction. The narrowness of this range indicates that the sample truly represents the traveling population in the sense that no day is over or underrepresented. The introduction of a higher sampling rate on Sunday means that those travelers are somewhat overrepresented. This bias is to some extent alleviated by the fact that the response rate was lower on the Sunday.

The second point that must be made about the data describing the train travelers concerns the response rate ("% response") figures. The "# of question-

Table 6-1
Summary of Rail Traffic and Survey Response

	No. of Passengers on Trains		% on Trains Surveyed	No. of Questionnaires Returned	% Response
	On All Trains	On Trains Surveyed			
Edinburgh-Glasgow					
Mon 22	4,137	792	26.3	585	73.9
Tues 23	3,752	1,075	37.3	716	66.6
Wed 24	3,866	1,014	34.9	588	58.0
Thurs 25	3,987	815	27.0	484	59.4
Fri 26	4,533	1,147	32.4	606	52.8
Sat 27	3,731	688	25.0	404	58.7
Sun 28	2,039	1,285	72.1	445	34.6
Total	26,045	6,816	34.3	3,828	56.16
Glasgow-Edinburgh					
Mon 22	4,621	890	28.5	657	73.8
Tues 23	4,281	931	31.0	664	71.3
Wed 24	4,491	854	28.1	547	64.0
Thurs 25	4,540	1,014	32.4	552	54.4
Fri 26	6,380	1,718	33.9	752	43.8
Sat 27	6,144	1,626	33.8	611	37.6
Sun 28	2,602	1,591	72.9	598	28.0
Total	33,059	8,624	35.4	4,381	50.8
Overall Total	59,104	15,440	34.9	8,209	53.17

naires returned" data was collected at the time of the survey, and it was not until the questionnaires were coded that it became obvious that some of the questionnaires had been filled in by people traveling to intermediate stations. To express the response rate as:

$$\frac{\text{number of questionnaires returned}}{\text{number of E-G passengers on trains surveyed}} \times 100$$

is to overstate the response rate. This becomes most obvious when the number of questionnaires returned exceeds the number of E-G passengers (denoted in the tables as "% response 100." In spite of this problem, some information as to the broad order of magnitude of the response rate and also as to its performance through time can be obtained. It will be noted that the rate tends to fall as the week progresses. It is thought that this may be because later in the week some travelers were passing through the system for the second time.

The data on the car-traveling population is presented in Tables 6-2 and 6-3. Both the weekday and the weekend traffic display normal characteristics with respect to average flow levels and peaking. Since the roadside survey was continuous, it was impractical to maintain statistics on the number of Edinburgh-Glasgow cars surveyed and the number of questionnaires distributed in a given period. Also, since the destinations of cars not surveyed were unknown, any attempt to calculate response rate figures would have been a meaningless

Table 6-2
Summary of Traffic Flows: By Hour

Time	Average Flow	Average Weekday Flow
07:00*	254 (136)#	331 (200)
08:00	411 (161)	621 (227)
09:00	304 (160)	371 (248)
10:00	298 (157)	311 (253)
11:00	295 (157)	250 (237)
12:00	272 (130)	235 (198)
13:00	299 (125)	284 (194)
14:00	343 (137)	342 (226)
15:00	346 (134)	376 (216)
16:00	439 (148)	514 (252)
17:00	511 (112)	648 (181)
18:00	425 (64)	429 (91)

* Hour Beginning
# Commercial

Table 6-3
Summary of Traffic Flows: By Day

Time	Total Traffic (12 hours)	Average Hourly Flow
Sat 13	3871 (783)	323 (65)
Sun 14	3592 (389)	299 (32)
Mon 15	4251 (1030)	354 (86)
Tues 16	4821 (2595)	402 (216)
Wed 17	4792 (2555)	399 (212)
Thurs 18	4874 (2642)	406 (220)
Fri 19	4722 (2451)	393 (204)
Sat 20	3673 (1018)	306 (84)
Sun 21	3220 (453)	268 (38)
Mon 22	4454 (2376)	371 (198)
Total	38,399 (16269)	

exercise. Moreover, since no detailed check was kept on the number of questionnaires distributed (mainly due to problems of bad weather, destroyed and lost questionnaires, etc.) only a crude response rate can be calculated. It is estimated that 13,000 questionnaires were distributed; 4,604 were finally coded (representing questionnaires returned with some, albeit minimal, response). The response rate can, therefore, be calculated as 35.4 percent. Although the measure is acknowledged to be a crude one, it clearly represents the correct order of magnitude. That this can be stated with confidence becomes clear if a range of error is considered. Even assuming that the estimate of the number of questionnaires distributed was wrong by 500 in either direction, the response rate would be within the range 34.1 to 36.8 percent. A response rate within this range can be regarded as successful for a postal return questionnaire of such complexity. It compares favorably with the train survey equivalent overall response rate of 53.17 percent.

Derivation of the Final Data Set

The number of questionnaires returned and coded contained a large number of responses that, for a number of reasons, were considered inadequate. Tables 6-4 and 6-5 illustrate the way in which the data set initially coded was eroded until the final data set emerged.

Given the nature of the model to be constructed, a questionnaire had to contain certain basic information in order to be usable for the purposes of analysis. The basic information required was:

Table 6-4
Derivation of Final Data Set: Train

Total Number of Questionnaires Coded =	7647	% of Total Coded
Reject 1: No Total Alternative Journey Cost	5081	66.4
Reject 2: No Total Actual Journey Cost	165	2.2
Reject 3: No Total Alternative Journey Time	80	1.0
Reject 4: No Income	118	1.5
Reject 5: No Total Actual Journey Time	44	0.6
Reject 6: Inappropriate Journeys	224	2.9
Reject 7: Alternative Mode = Bus	594	7.7
Reject 8: Insufficient Walk/Wait Times	61	0.8
Final Data Set =	1280	16.74

Table 6-5
Derivation of Final Data Set: Car

Total Number of Questionnaires Coded =	4604	% of Total Coded
Reject 1: No Total Alternative Journey Cost	2762	60.0
Reject 2: No Total Alternative Journey Time	251	5.1
Reject 3: No Total Journey Cost	63	1.4
Reject 4: No Total Journey Time	32	0.7
Reject 5: No Income	27	0.6
Reject 6: Inappropriate Journeys	19	0.4
Reject 7: Alternative Mode = Bus	168	3.6
Reject 8: Insufficient Walk/Wait Times	16	0.3
	1266	27.5

1. Total Cost of Journey by Alternative Mode
2. Total Cost of Journey by Chosen Mode
3. Total Time of Journey by Alternative Mode
4. Total Time of Journey by Chosen Mode
5. Income.

At a later stage, it was decided that information on "walk/wait" times should be added to this list:

6. Walk/Wait Times.

Having decided to carry out the first analysis using only those travelers with a car vs. train choice, those with a bus alternative mode were also eliminated.

Finally, some journeys, notably those involving boats to the West Coast islands or airplanes, were judged to be so unusual as to be a disturbing influence, and such journeys were eliminated, together with a small number of journeys to intermediate stations, which had failed to be eliminated at an earlier stage.

Clearly, the most striking feature of the results of this elimination procedure is the very high number of responses rejected for failing to provide information on the "Total Alternative Journey Cost," 66.4 percent and 60 percent for train and car, respectively. Although rejection on these grounds does not imply that all the other information sought was complete, and although the numbers rejected at this stage actually include many minimal responses, the lack of this particular piece of information is important and should be kept in mind when testing of the model is undertaken.

The final sample of train travelers is 16.74 percent of the train responses coded, 8.3 percent of the total number of Edinburgh-Glasgow train passengers surveyed, and 2.9 percent of the total population of passengers between Edinburgh and Glasgow during the period of the survey. The final sample of car travelers is 27.5 percent of the car responses coded; given that the exact number of questionnaires distributed and the sampling ratio are not known, the calculation of the sample percentage for car travelers becomes somewhat speculative. However, assuming 13,000 questionnaires distributed and a 75 percent "car stopping rate," it is estimated that the total population of Edinburgh-Glasgow travelers through the survey point was 17,319, yielding the result that the final data set is 7.3 percent of the total population surveyed. A *caveat* should be added at this point: the apparently low total sample figure for the train is misleading in the sense that the passenger counts are based on individual travelers, whereas the questionnaires represent parties of travelers. The 2.9 percent rate is, therefore, a considerable understatement of the true rate.

Comparison of Initital and Final Data Sets

It would be unreasonable to present a reduction in the size of the data set of the magnitude described in the preceding section without considering the biases that may occur as a result of such a procedure. A number of variables have been selected, notably:

1. Journey purpose
2. Journey frequency
3. Income
4. Age
5. Sex.

It is felt that those variables are representative of the sample and that an examination of the biases in these variables resulting from the reduction in size

Table 6-6
Comparison of Final and Initial Data Sets: Journey Purpose

Purpose	Initial Set (12251)		Final Set (2546)	
	#	%	%	#
Social-Recreational	4617	37.7	26.9	685
Journey to/from Work	3068	25.0	35.9	915
Firm's Business	1389	11.3	14.7	374
Personal Business	1246	10.2	8.8	224
Shopping	375	3.1	2.2	57
Educational	581	4.7	4.8	123
Other (including Multipurpose Journeys)	975	8.0	6.6	168
Journey to/from Work	3068	25.0	35.9	915
Firm's Business	1389	11.3	14.7	374
All Other Purposes	7694	62.8	49.4	1257

Table 6-7
Comparison of Final and Initial Data Sets: Frequency

Frequency	Initial Set		Final Set	
	#	%	%	#
Daily	1349	11.0	12.9	329
2, 3, 4 Days/Week	861	7.0	11.5	292
Once a Week	1117	9.1	10.2	261
1, 2, 3 Times/Month	2172	17.7	22.5	574
Less Than Once a Month	6484	52.9	42.3	1077
(No Response)	(262)	(2.2)	(0.5)	(13)

Correlation Coefficient $(\%_I, \%_F) = 0.313$

of the data set will clearly indicate the magnitude of the bias problem, should it exist.

Tables 6-6 through 6-10 show the relationships between both the absolute numbers and the percentages of travelers in each category for the initial and final data sets. (The train and car data sets have been combined for this exercise.)

As for the journey purpose comparison, it is clear that the journey-to-work travelers are overrepresented in the final data set, having exchanged representation with the social-recreational travelers. However, the problem is somewhat mitigated by the data of Table 6-6. Since the aim of this study was basically to examine trips that were neither journeys to work nor business trips, all other trips have been aggregated, and it is felt that although social-recreational trips are still underrepresented in the final data set, the use of all purposes (other than business and journey to work) as an aggregated category makes the problem less acute. By way of explanation, it might be said that this result is not unexpected.

Table 6-8
Comparison of Final and Initial Data Sets: Income

Income	Initial Set		Final Set	
	#	%	%	#
< £ 100 p.a.	3704	30.2	17.9	456
£ 1001 − £ 1250 p.a.	1722	14.1	12.4	3.6
£ 1251 − £ 1500 p.a.	1282	10.5	10.8	276
£ 1501 − £ 1750 p.a.	936	7.6	10.6	269
£ 1751 − £ 2000 p.a.	806	6.6	9.6	244
£ 2001 − £ 2250 p.a.	609	5.0	7.5	191
£ 2251 − £ 2500 p.a.	461	3.8	5.8	148
> £ 2501 p.a.	1853	15.1	25.4	646
(No Response)	(878)	(7.2)	−	−

Correlation Coefficient ($\%_I, \%_F$) = 0.169

Table 6-9
Comparison of Final and Initial Data Sets: Age

Age	Initial Set		Final Set	
	#	%	%	#
Under 20 yrs. old	1612	13.2	7.0	179
21-30	3513	28.7	30.3	772
31-40	2401	19.6	24.9	633
41-50	2205	18.0	20.1	534
51-60	1560	12.7	12.7	324
61 yrs. and over	748	6.1	3.8	96
(No Response)	(212)	(1.7)	(0.3)	(8)

Correlation Coefficient ($\%_I, \%_F$) = 0.727

Table 6-10
Comparison of Final and Initial Data Sets: Sex

Sex	Initial Set		Final Set	
	#	%	%	#
Male	9262	75.6	84.9	2161
Female	2958	24.1	14.6	371
(No Response)	(31)	(0.2)	−	−

Given the complex and detailed nature of the information elicited by the questionnaire, it is not unreasonable to find that a lower percentage of the "casual" as opposed to the "regular" travelers provided full information. A businessman or a person making a regular journey to work would seem more likely to possess full information about his trip than a person traveling for social or recreational purposes.

This contention is substantiated by an examination of Table 6-7, which shows that the infrequent travelers are underrepresented in the final data set. In this case, a correlation coefficient was computed to examine the relationship between the percentages in each category in the initial and final data sets and was found to be 0.313. While this cannot be taken to represent a strong relationship between the two data sets, it is not too discouraging.

Unfortunately, the same cannot be said for the next variable under consideration: income, with a correlation coefficient of 0.169. However, inspection of the data reveals a clear pattern of underrepresentation of the lower-income groups and increasing overrepresentation by the higher-income groups. While such a situation cannot be held to be particularly satisfying, some comfort can be found in the observation that it was predictable and in line with the findings of other studies.

Examination of the age variable reveals an equally predictable underrepresentation in the lowest and highest age groups, but the data as a whole supports a correlation coefficient of 0.727; both factors would lead to satisfaction with this aspect of the final data set.

Finally, it remains to note that females are underrepresented in the final data set. No explanation is offered for this phenomenon.

In conclusion, it might be said that in designing a survey of this nature, it must be realized that the subjects will differ both in the amount of information they possess and in their willingness to transmit it in an interview situation. The biases observed in the final data set are the result of predictable differences in the ability and willingness of the subjects to respond to a complex questionnaire. The biases met are endemic to this type of survey research and, as such, need not restrict the confidence with which the subsequent steps of the analysis can be undertaken.

"Internal" Checks on the Data

Before proceeding to any analysis of this data, it is advisable to confirm that the data possesses certain characteristics that may be required by the statistical techniques used at a later stage. The most important characteristics concern the distributions of the variables and the relationships between them and can be examined by considering their frequency distributions and correlation matrix. To be forewarned of any peculiarities in the data is to be forearmed when dealing with the statistics.

To present a frequency distribution of every variable would be time-consuming and tedious. However, the twelve major variables were selected and their distributions checked for normality. Only two points must be made. First, the distributions of the time and cost difference variables are very close to the bell shape of a normal distribution, which implies that they are unlikely to cause problems in the statistical estimation of the model. Second, the distributions of the variables not representing differences all appear to be basically bell-shaped, but all are skewed to the right. This is because the range of possible times and costs are virtually unbounded above the mean, whereas below the mean it is constrained by the physical limitations on the time necessary for a journey and the institutional limitations on its cost. These distributions represent variables that behave as expected and, therefore, can be used with confidence to estimate the model.

The matrix of correlation for the sixteen variables considered most representative of those available in the final data set is presented in Table 6-11. It is encouraging to note that very few of the coefficients are high; given the large size of the sample, only coefficients greater than 0.70 are considered to be troublesome. Of the six coefficients whose values are greater than 0.70, two represent correlations between a "difference" variable and one of the two components used to calculate the difference, two represent correlations between time and cost differences and "line-haul" time and cost difference and total transit time difference, and the final one represents the correlation between total transit time difference and "line-haul" time difference. In all six cases, the high correlation coefficient does not pose a real problem, since the correlated variables will never appear together in the models to be tested.

Checks for "Interurban-ness"

The aim of this study is to build a mode-choice model in an interurban situation, and it was asserted that a journey between Edinburgh and Glasgow constituted an interurban trip. While from a knowledge of the locality it is clear that such a trip is truly interurban, it may be difficult, in a different locality and cultural situation, to conceive of such a short journey being interurban. Moreover, it may be useful to have a criterion for defining an interurban trip should similar studies be contemplated.

Two factors should be taken into account: the frequency with which the trip is undertaken and the amount of commuting that takes place on the route. In the Edinburgh-Glasgow case, 52.9 percent of all respondents reported that they made the journey "less than once per month," and only 25 percent of all trips were commuting trips. It is not intended, in this study, to construct an index of "interurban-ness," although this might be an interesting exercise. It is contended, however, that the figures quoted above demonstrate that the journey under consideration is clearly an interurban one.

Table 6-11
Matrix of Correlation Coefficients

		1	2	3	4	5	6	7	8	9	10	11	12	13	14	15	16
CHOICE	1	1.000															
TIMDIF	2	-.122	1.000														
COSDIF	3	-.047	.080	1.000													
JU DIF	4	-.105	.188	.030	1.000												
SUBTIM	5	-.126	.405	.058	.273	1.000											
TJT-TR	6	-.003	.629	.084	.247	.570	1.000										
TJC-TR	7	.014	.141	.717	.069	.177	.378	1.000									
CU-TRA	8	-.148	.242	.040	.849	.316	.290	.082	1.000								
INTRAN	9	-.126	.450	.058	.273	1.000	.570	.177	.316	1.000							
SUBCOS	10	-.099	-.051	.033	-.052	.285	.085	.109	-.063	.285	1.000						
WW-TIM	11	-.220	.379	-.024	.164	.255	.335	-.049	.217	.255	-.061	1.000					
TJT-CA	12	.117	-.201	.026	.124	.316	.636	.336	.125	.316	.157	.046	1.000				
TJC-CA	13	.082	.067	-.474	.046	.143	.361	.274	.048	.143	.091	-.029	.388	1.000			
TRANSD	14	.032	-.908	-.097	-.129	-.322	-.527	-.174	-.163	-.322	.027	.045	.238	-.086	1.000		
LHTIMD	15	-.038	-.703	-.067	.021	.234	-.219	-.079	.011	.234	.189	.190	.422	-.007	.845	1.000	
LHCOSD	16	-.024	-.096	-.783	-.057	.130	-.015	-.509	-.071	.130	.595	-.019	.077	.438	.095	.171	1.000

7

The Method of Analysis

The models of economists are frequently estimated using the techniques of multiple-regression analysis. However, for a number of reasons the adequacy of regression analysis for the purposes of this study is in doubt. The aim of this chapter is to delineate these inadequacies and consider a number of alternative techniques, notably probit analysis, logit analysis, and discriminant analysis, which may be better suited to the type of analysis being undertaken.

Regression Analysis

The problems associated with the use of regression analysis stem from the nature of the phenomenon to be explained, i.e., the choice of either train or car. The binary choice problem appears on a number of occasions in economics, typically in analyses of housing (buy vs. rent) and consumer durables (buy a new car vs. not buy a new car; own a dishwasher vs. not own a dishwasher).

In such cases, the dependent variable can be assigned one of two values:

$$y_i = \begin{cases} 1; \text{ if the } i\text{th person chooses the train} \\ 0; \text{ if the } i\text{th person chooses the car} \end{cases} \tag{7-1}$$

The simplest formulation utilizes the *linear probability function* and computes least-squares estimates of the coefficients of the model:

$$y = X\beta + \epsilon \tag{7-2}$$

Thus, the expected value of y is a linear combination of the explanatory variables. Given the classical least-squares assumptions:

$$E(\epsilon) = 0$$
$$\text{and} \tag{7-3}$$
$$E(\epsilon^2) = \sigma^2 I$$

the classical least-squares estimates of the β's are obtained, and

$$E(y|X) = X\beta \tag{7-4}$$

the conditional expectation of y given the X's, results. Given the 0,1 nature of the dependent variable, the conditional expectation of y can be interpreted as the conditional probability of the occurrence of the event (in this case, the choice of train as the travel mode) given the values of the explanatory variables.

This linear probability function has associated with it three problems: heteroskedasticity; non-normal distribution of ϵ_i; and unbounded predictions.

Heteroskedasticity

The 0,1 form of the dependent variable leads naturally to restrictions of the values that can be taken by the disturbance term, ϵ_i. From (7-2),

$$\epsilon = y - X\beta \tag{7-5}$$

and for a given set of X's, X'_t, i.e., the row X' at time t,

$$\epsilon_t = y_t - X'_t\beta \tag{7-6}$$

Given that y_t can only take values of 0 or 1, ϵ_t must be equal to either $-X'_t\beta$ or $1-X'_t\beta$. Then, in order for $E(\epsilon_i) = 0$ to hold, the distribution of ϵ_i must be:

$$
\begin{array}{cc}
\epsilon_t & f(\epsilon_t) \\
\hline
-X'_t\beta & 1 - X'_t\beta \\
1 - X'_t\beta & X'_t\beta
\end{array}
\tag{7-7}
$$

Thus, the variance of ϵ_i is:

$$\mathrm{VAR}(\epsilon_t) = E(\epsilon_t^2) = (X'_t\beta)(1 - X'_t\beta) \tag{7-8}$$

In other words, the variance of ϵ is shown to depend upon the X's and, therefore, is not constant, $\sigma^2 I$.

This violates the classical least-squares assumption of homoskedasticity, with the result that the estimates of β, although linear and unbiased, are not efficient, in the sense that of all the linear unbiased estimates of β, they are not minimum variance.

BLUE estimators can be obtained by using the variance-covariance matrix:

$$
E(\epsilon^2) =
\begin{bmatrix}
X'_1\beta(1 - X'_1\beta) \cdots \cdots \cdots \cdots \cdots 0 \\
0 \quad X'_2\beta(1 - X'_2\beta) \\
\vdots \\
\vdots \\
\vdots \\
0 \cdots \cdots \cdots \cdots \cdots X'_T\beta(1 - X'_T\beta)
\end{bmatrix}
= \Omega
\tag{7-9}
$$

in the generalized least-squares formula

$$\beta = (X'\Omega^{-1}X)^{-1}X'\Omega^{-1}y \tag{7-10}$$

However, Ω is unknown; Goldberger suggests estimating it by using classical least squares estimates of β in (7-9). Thus:

$$\Omega^* = \begin{bmatrix} \hat{y}_1(1 - \hat{y}_1) & \cdots & 0 \\ \vdots & & \vdots \\ 0 & \cdots & \hat{y}_T(1 - \hat{y}_T) \end{bmatrix} \tag{7-11}$$

is an estimate of the error variance-covariance matrix Ω based on $\hat{y}$'s derived from a classical least-square regression. It is argued that since $\hat{y}_t$ is an unbiased estimate of $E(y_t)$, Ω^* may be used as an estimate of Ω. However, Goldberger recognizes that: "To be sure, b^* is not the BLUE of β; but it does take account of heteroskedasticity."[1] Such a solution is less than satisfactory.

Non-Normal Distribution of ϵ_i

Apart from the estimation problems resulting from the unusual distribution of the ϵ_i, a further problem arises. Because the ϵ_i are not normally distributed, the estimators of β are also not normally distributed. The asymptotic means and variances of β can, however, be determined without difficulty; the asymptotic means are equal to the true values, and the asymptotic variance can be readily determined.[2] The problem arises from the fact that given the non-normality of the distributions of the estimators, the classical tests of significance do not apply. Tests of significance on the β's must proceed by deriving the acceptance region from the known distribution of ϵ_i. This is likely to be a difficult process. The same conclusions apply to the tests of significance of the regression as a whole, for example, the F-test. The fact that ϵ_i is not normally distributed means that no confidence can be placed on the F-values computed on the regression.

Unbounded Predictions

The final and most intractable problem associated with the use of binary dependent variables in regression analysis concerns the predictions from the model. As was noted above, the conditional expectation of y can be interpreted as the conditional probability of the occurrence of the event being considered, given the X's. This implies, however, that

$$0 \leq E(y|X) \leq 1 \qquad\qquad (7\text{--}12)$$

The prediction of $y, \hat{y}$, does not meet this restriction since, as a point on a straight line, it can take values from $-\infty + \infty$.

Summary

Given that the use of a dichotomous dependent variable leads to estimates which are not *BLUE*, which cannot be tested for significance, and which may produce predictions outside the 0,1 range which vitiate the probability interpretation of the model, it must be concluded that the suitability of least-squares regression for the analysis of this type of problem is in doubt. While the effect of the first two objections on the properties and significance may be mitigated by appealing to the asymptotic properties of estimators based on large samples, the problem of the unbounded predictions remains intractable and is a real difficulty in cases where predictions are an integral part of the analysis.

It is, then, appropriate to consider other methods of analysis which may be more appropriate. Two approaches based on dichotomous dependent variables and the necessity to restrict the range of the $E(y)$ are probit and logit analysis. ("Tobit" analysis will not be considered as its applications are appropriate in cases where only a single bound exists, i.e., $0 \leq E(y \mid X) \leq \infty$.) Both methods proceed by creating a transformation of the linear combination, $X\beta$, such that, while X may take values from $-\infty$ to $+\infty$, the dependent variable will be limited to the range 0,1.

Probit Analysis

Probit analysis has a long history in biometrics, having been developed by Finney (19) for the analysis of toxicology problems. For example, it was often desired to relate the occurrence of an event, e.g. the death of an insect, to the dosage of insecticide the insect had received. In short, probit analysis is used to determine the relationship between the probability that an insect will be killed and the strength of the dose of poison administered, where the dependent variable is clearly dichotomous: killed or not killed. The key concept in probit analysis, whether used for biological assay or economic analysis, is that of a threshold level of the explanatory variable (assuming for simplicity a single explanatory variable). In the insecticide problem, it is assumed that there is a threshold dosage level, above which the insect will be killed and below which it will live. In economic applications, the same concept holds. For example, a threshold level of income is assumed to be required before a household will buy a new car.

In both types of application, the threshold values are assumed to be normally distributed over the population as a result of which the parameters of the distribution are estimated using maximum likelihood methods from data that show the number in the sample observed to be in one or other category (own/not own; killed/not killed) at various levels of the explanatory variables (income; dosage). Clearly, for most economic applications, more than one independent variable may be necessary, so that the use of multivariate probit analysis becomes necessary.

The Probit Model[3]

The dependent variable W is postulated to take the values 0 or 1, depending upon the values of the independent variables, $X_1 \ldots \ldots X_m$. An index, I, is then constructed which is a linear combination of the independent variables.

$$I = \beta_0 + \beta_1 X_1 + \beta_2 X_2 + \cdots \beta_m X_m \qquad (7\text{--}13)$$

The concept of a linear relationship is analogous to the linear relationship in regression analysis and may be circumvented if necessary by transformations of the independent variables to give a relationship which, while linear in the parameters, is nonlinear in the variables.

If I_i is the value of I evaluated from (7-13) using the values of X corresponding to the ith household, and $\bar{I}_i$ is the critical, or threshold, value of $\bar{I}$ for the ith household, then

$$W_i = 1 \text{ for } I_i \geq \bar{I}_i$$
$$\text{and} \qquad\qquad (7\text{--}14)$$
$$W_i = 0 \text{ for } I_i < \bar{I}_i$$

In other words, if the value of I_i calculated from (7-13) is equal to or greater than the critical value, $\bar{I}_i$, then $W_i = 1$; if I_i is less than $\bar{I}_i$, then $W_i = 0$.

The $\bar{I}_i$'s, or critical values, are assumed to be normally distributed over the population, $N(0,1)$. The fact that the $\bar{I}_i$'s are normally distributed represents differences among households which are either random or the result of variables not included in the model.

Given (7-14), the probability that, given I, W_i will equal 1 is:

$$P(W = 1 \mid I) = P(\bar{I}_i \leq I_i) = P(I) = \frac{1}{\sqrt{2\pi}} \int_{-\infty}^{I} \exp(-\tfrac{1}{2}u^2) du$$

$$(7\text{--}15)$$

Therefore, the probability that, given I, W_i will equal 0 is:

$$P(W = 0|I) = P(\bar{I}_i > I_i) = 1 - P(I) = Q(I) = \frac{1}{\sqrt{2\pi}} \int_I^\infty \exp(-\tfrac{1}{2}u^2)du$$

$$(7\text{--}16)$$

Thus, I is the probit of $P(W)$ and is defined as the abscissa which corresponds to a probability $P(W)$ in a normal distribution with mean, 0, and variance, 1.

The parameters of the model can be estimated by maximum likelihood methods.

Consider sample of observations on m variables at s points $(X_{ij}$; $i = 1 \ldots m$; $j = 1 \ldots s$). Let n_j be the total number of observations at the jth point; let r_j be the number of observations at the jth point for which $W = 1$; then $n_j - r_j$ = the number of observations at the jth point for which $W = 0$. The likelihood of the sample is a function of the values $(b_0, b_1 \ldots b_m)$ assumed to represent the population parameters $(\beta_0, \beta_1 \ldots \beta_m)$:

$$L(b_0, b_1 \cdots b_m) = \prod_{j=1}^s [P(b_0 + b_1 X_{1j} + \cdots + b_m X_{mj})]^{r_j}$$

$$[Q(b_0 + b_1 X_{1j} + \cdots + b_m X_{mj})]^{n_j - r_j}$$

Recall that, (7–15)

$$P(W) = \frac{1}{\sqrt{2\pi}} \int_{-\infty}^I e^{\left(\frac{-u^2}{2}\right)} du \qquad (7\text{--}17)$$

and $Q(W) = 1 - P(W)$

Now let

$$Y_j = b_0 + b_1 X_{1j} + \ldots b_m X_{mj}; \; P_j = P(Y_j); \; Q_j = Q(Y_j) \quad (7\text{--}18)$$

To find the maximum likelihood estimates of the population parameters, it is convenient to find the value of the b's that maximize $\log L$ (rather than L).

$$L^*(b_0, b_1 \cdots b_m) = \text{Log } L(b_0, b_1 \cdots b_m)$$

$$= \sum_{j=1}^s (r_j)\log P_j + (n_j - r_j) \log Q_j$$

$$(7\text{--}19)$$

Thus, the conditions for a maximum are the $m + 1$ equations that result from finding the partial derivatives of L^* and setting them equal to zero. These equations can be shown to be:

$$L_i*(b_0, b_1 \cdots b_m) = \sum_{j=1}^{s} r_j \frac{X_{ij}Z_i}{P_j} - (n_j - r_j)\frac{X_{ij}Z_i}{Q_j} = 0$$

$$(i = 0, 1, 2 \cdots m)$$

(7–20)

The equations can be solved by an iterative process. As the algebra is complex and tedious, it is not intended to present the solution here.[4]

Testing the Estimates for Significance

Hypotheses about the β's may be tested by the likelihood ratio method. In short, this consists of setting up a hypothesis about the estimators and evaluating the likelihood function with and without the restrictions on the hypothesis. The ratio of the two likelihoods is the key to the significance test.

Consider the hypothesis that $P(W = 1)$ is independent of the X's. This probability is given by:

$$P(w = 1) = P(\bar{I}_i \leq \beta_0) = P(\beta_0) = \frac{1}{\sqrt{2\pi}} \int_{-\infty}^{\beta_0} e^{\left(\frac{-u^2}{2}\right)} du$$

(7–21)

If the hypothesis is true, the maximum likelihood estimator of β_0 is the value of b_0, which maximizes

$$L(b_0, 0 \cdots 0) = [P(b_0)]^r [Q(b_0)]^{n-r}$$

$$\text{where } r = \sum_{j=1}^{s} r_j \text{ and } n = \sum_{j=1}^{s} n_j$$

(7–22)

Thus, the value b_0 which maximizes (7-20) is found to be such that:

$$P(b_0') = \frac{r}{n}$$

(7–23)

and the value of the log of the likelihood function evaluated for the maximum likelihood estimate of β_0 is:

$$L*(b_0, 0 \cdots 0) = r\left(\log \frac{r}{n}\right) + (n - r)\log \frac{n - r}{n}$$

(7–24)

Removing the restriction of the hypothesis, L^* is obtained from (7-17) using the P_j and Q_j, which correspond to the maximum likelihood estimators of the b's.

Now let:

$$\log \lambda = L^*(b'_0, 0 \cdot \cdot \cdot 0) - L^*(b_0, b_1 \cdot \cdot \cdot b_m) \qquad (7\text{--}25)$$

then $-2 \log \lambda$ is distributed approximately like χ^2 with m degrees of freedom for large samples when the hypothesis is true.

Other tests of significance regarding the β's can be performed, for example, on the hypothesis that $\beta_k = 0$ or $\beta_2 = \beta_4$, etc. However, since each test requires considerable computation, it is useful to make use of the approximate normality of the distribution of maximum likelihood estimators from large samples. The b_k are approximately distributed by the ($m + 1$)-variate joint normal distribution with means β_k and estimated variance-covariance matrix $\| - L_{ik} \|^{-1}$. Thus, for example, the t-test can be used to test the hypothesis that $\beta_0 = \beta_1 = \ldots \beta_m = 0$.

Interpretation of the Probit Model

In some cases the problem of choosing between least-squares regression and probit analysis has been made on the grounds that the regression coefficients can be readily interpreted, whereas the probit coefficients pose greater problems. In the light of this misconception it is appropriate at this point to say something about the interpretation of the probit coefficients.

The interpretation is made less difficult if it is recalled that the dependent variable in the probit relationship is not I, the linear combination of the independent variables, but $P(X)$, its unit cumulative normal transform. Thus, the probit coefficients can be explained in the following way: a one-unit change in X_1 will produce a change of β_1 standard deviation units on the probability, $P(X)$. The constant term β_0 indicates the number of standard deviations from the 50:50 point (equal probability of the occurrence or nonoccurrence of the event) when all the independent variables are zero.

An example [from Lisco (42)] will clarify the exposition. Assume that the following relationship is observed:

$$P(X) = -.5 + 1.2X \qquad (7\text{--}26)$$

This may be interpreted as saying that when $X = 0, P(X)$ will take a value .5 standard deviations below the center of the distribution (the 50:50 point), i.e., $P(X) = .69$. In other words, a randomly selected individual with a zero value of X would have a .69 probability that the event would occur, in Lisco's example of choosing mode 1 and rejecting mode 2. Positive values of X would change the probability by 1.2 standard deviations for each unit change in X.

Summary

Probit analysis has been shown to avoid the problem of unbounded predictions by utilizing a transformation bounded by zero or one. The relationships estimated can be tested using standard statistical tests of significance, and meaningful interpretations can be made of the maximum likelihood estimates of the coefficients. It should be stressed, however, that the probit method is dependent upon the assumption that the threshold values of the linear combination of independent variables are normally distributed over the population. It has been argued that this assumption is unnecessarily restrictive. In fact, the properties of the model hold if a weaker assumption can be made, notably that there exists a transformation of the threshold values which is normally distributed. Discussions of transformations of an unknown and unobservable distribution, however, are somewhat meaningless, and the assumption of normality may be accepted if it is interpreted as representing factors that have not been incorporated into the model.

Logit Analysis

Like probit analysis, logit analysis originated in the field of biometrics and is similarly aimed at a solution to the problem of restricting estimates of probabilities within the bounds of 0 and 1. The basic approach of the two methods is the same: to find a transformation of the probability which can take values from $-\infty$ to $+\infty$, while restricting the probability itself to values in the range 0 to 1. This similarity means that the logit procedure can be presented in a somewhat abbreviated form.

The Logit Transformation

Following an example from Stopher (69), suppose that a traveler has a choice between car and mass transit, with a probability p that he will use the car and hence $1 - p$ that he will use mass transit. Suppose, further, that his choice of mode is influenced by the differences in time and cost, Δt and Δc, respectively, between the modes. Then, the linear probability function approach leads to:

$$p = \beta_0 + \beta_1(\Delta_t) + \beta_2(\Delta c) \qquad (7\text{--}27)$$

This is the objectionable formulation which may lead to unbounded predictions of the probabilities. Consider, then, the ratio of p to $1 - p$, the odds in favor of a positive response (given the values of the explanatory variables). As the probability increases from zero to one, the odds increase from zero to infinity; moreover, as the probability increases from zero to one, the natural logarithm of the odds, $\log \dfrac{p}{1 - p}$, increases from minus infinity to plus infinity. The natural logarithm of the odds is known as the logit of a positive response. Figure 7-1 shows the logit as a function of the probability.

It is now postulated that the logit is a linear combination of the explanatory variables.

$$\log \left[\frac{p}{1 - p} \right] = X\beta \tag{7-28}$$

where

$$X\beta = \beta_0 + \beta_1 X_1 + \beta_2 X_2 + \ldots + \beta_n X_n \tag{7-29}$$

Then:

$$\frac{p}{1 - p} = e^{X\beta} \tag{7-30}$$

and

$$p = \frac{e^{X\beta}}{1 + e^{X\beta}} \tag{7-31}$$

Substituting for $X\beta$

$$p = \frac{e^{(\beta_0 + \beta_1 X_1 + \beta_2 X_2 + \ldots + \beta_n X_n)}}{1 + e^{(\beta_0 + \beta_1 X_1 + \beta_2 X_2 + \ldots + \beta_n X_n)}} \tag{7-32}$$

In order to facilitate the following discussion, it is appropriate to consider the probability q where $q = 1 - p$. In the mode-choice example, if p is the probability of choosing the train, then q is the probability of choosing the car and

$$q = \frac{1}{1 + e^{X\beta}} \tag{7-33}$$

Thus, the values that $X\beta$ can take are unbounded, but the values of q are restricted to values between zero and one, since the function in (7-31) is a special case of the more general logistic function

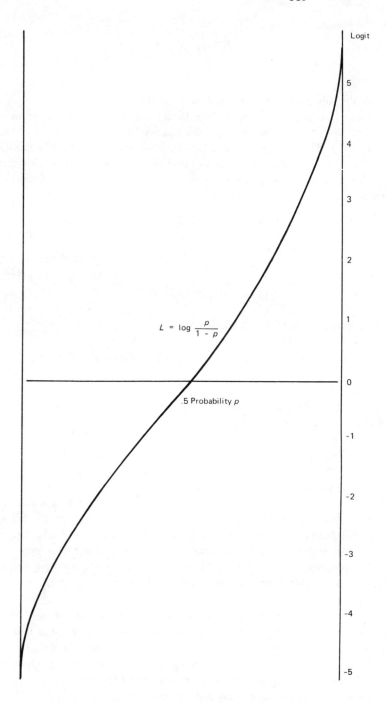

Figure 7-1. The Logit as a Function of the Probability

$$q = \frac{\gamma}{1 + e^{X\beta}} \qquad (7\text{-}34)$$

in which the values of q are restricted between 0 and γ.[5]

Consistent estimators of the β's can be derived by a process analogous to the derivation of the estimators in the above section on probit analysis (pp. 103-105). The data is grouped on X. Then let $n(x)$ be the number of observations at X, let $p(x)$ be the number of those observations for which $q = 1$, and let $q(x)$ be the number for which $q = 0$. Then the likelihood of the sample is

$$L = \prod [P(x)^{p(x)} Q(x)^{q(x)}]^{n(x)} \qquad \begin{aligned} P(x) &= \text{probability} \\ &\text{that } q = 1, \text{ given } X \end{aligned}$$

$$\begin{aligned} Q(x) &= \text{probability} \\ &\text{that } q = 0, \text{ given } X. \end{aligned}$$

$$(7\text{-}35)$$

and the log-likelihood of the sample is

$$L = \log \overline{L} = \sum [n(x)p(x) \log P(x) + n(x)q(x) \log Q(x)] \qquad (7\text{-}36)$$

where the sum is over the groups. Setting the differentials of $\overline{L}$ with respect to β equal to zero gives a set of nonlinear equations

$$\sum n(x)p(x)x = \sum n(x)P(x)x \qquad (7\text{-}37)$$

which can be solved iteratively.

It should be noted that this formulation of the logistic function does not include a stochastic error term of the type: $X\beta + \epsilon$. Rather, it utilizes the Bernouilli schemat, where the dependent variable, q, is made up of an unobservable probability plus an unobservable error term. This procedure is formally analogous to the probit estimation procedure presented above and should be contrasted with the regression analysis approach with its explicit stochastic error term.

Clearly, the logit transformation is similar to the probit analysis. The expression is almost identical to the cumulative normal curve, since it takes the form of a symmetrical sigmoid curve and differs from the normal sigmoid curve at the extremes (Figure 7-2). The parameters of the logistic function are somewhat easier to interpret than those of the probit function, since they do not involve the necessity to utilize standard deviation units.

At this point, it is important to note that the curve derived in this section

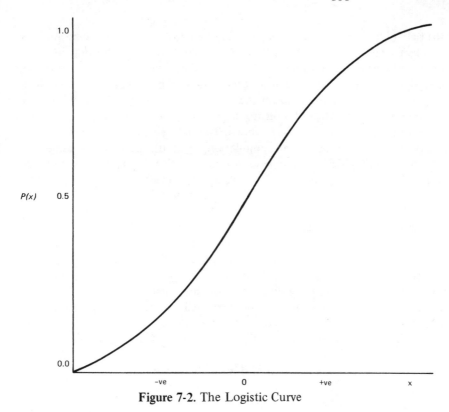

Figure 7-2. The Logistic Curve

conforms to the specifications of the curve derived on *a priori* grounds in Figure 2-5.

Summary

It is clear, then, that logit analysis is analogous to probit analysis in the sense that a transformation is used to avoid the problem of unbounded predictions. The transformations produce function relationships which are very similar, although it is argued that the logit formulation is conceptually more acceptable and, given the greater simplicity of the logit function, easier to interpret. It should be noted, however, that logit analysis requires an appeal to the asymptotic properties of large samples in order to produce consistent estimators.

Discriminant Analysis

The final method of analysis to be considered is discriminant analysis. Like probit analysis and logit analysis, discriminant analysis was developed for use in the field of biology, notably to enable biologists to identify subspecies. A typical problem would be the classification of a given flower as belonging to one or other subspecies when this could not be done simply by inspection. The underlying concept is that a linear combination of measurements on the flower can be found, such that the flower can be classified according to the value of the linear combination. The criterion for selection of the "best" linear combination is the minimization of misclassifications. The nature of a mode- or route-choice problem, with two observable populations, and the problem of assigning a newcomer, with given measurements, to one or other population, led to the use of discriminant analysis in the field of transport analysis.

The Discriminant Function

Discriminant analysis begins with the premise that the population from which the same is drawn is, in fact, made up of two subpopulations: car and bus users. Two assumptions about these subpopulations are usual:

1. The distributions of the variables for which measurements are obtained are multivariate normal.

2. The variance-covariance matrices of the variables are the same for both populations.

Now Z is defined as a function of the variables such that:

$$Z_{ij} = \lambda_1 X_{1ij} + \lambda_2 X_{2ij} + \ldots + \lambda_k X_{kij} \qquad (7\text{--}38)$$

where λ_p and X_{pij} are the weighting coefficient and the variable value for the pth factor ($p = 1 \ldots k$), of the jth person ($j = 1 \ldots n$) in the ith mode ($i = 1, 2$).

or:

$$Z_{ij} = \sum_{p=1}^{k} \lambda_p X_{pij} \qquad (7\text{--}39)$$

Discriminant analysis aims to find the values of λ_p that best discriminate between the two populations. The function that "best" achieves this end is one that provides the greatest separation between the two populations (e.g. the greatest distance between their means) relative to the variance within each subpopulation. Thus, the criterion is to select the weights, λ_p, such that the between-population variance is maximized relative to the within-population variance.

The between-population variance is the square of the distance between the population means:

$$(\overline{Z}_1 - \overline{Z})_2{}^2 = \left[\sum_{p=1}^{k} \lambda_p (\overline{X}_p{}^1 - \overline{X}_{p2}) \right]^2 \qquad (7\text{-}40)$$

and the within-population variance is:

$$\sum_{p=1}^{k} \sum_{q=1}^{k} \lambda_p \lambda_q C_{pq} \qquad (7\text{-}41)$$

where C_{pq} is the common variance-covariance matrix.

Thus, to maximize the between-population variance relative to the within-population variance, it is necessary to maximize a function, G:

$$G = \frac{\left[\sum_{p=1}^{k} \lambda_p (\overline{X}_{p1} - \overline{X}_{p2}) \right]^2}{\sum_{p=1}^{k} \sum_{q=1}^{k} \lambda_p \lambda_q C_{pq}} \qquad (7\text{-}42)$$

Differentiating with respect to λ_p, ($p = 1 \ldots k$), and setting $\dfrac{\partial G}{\partial \lambda_{\hat{p}}} = 0$ gives:

$$\sum_{q=1}^{k} \lambda_q C_{pq} = \frac{\sum_{p=1}^{k} \sum_{q=1}^{k} \lambda_p \lambda_q C_{pq}}{\sum_{p=1}^{k} \lambda_p (\overline{X}_{p1} - \overline{X}_{p2})} (\overline{X}_{p1} - \overline{X}_{p2}) \qquad (7\text{-}43)$$

Multiplying through by $C_{pq}{}^{-1}$, and summing over p gives, at a maximum:

$$\lambda_q = K \sum_{p=1}^{k} C_{pq}{}^{-1}(\overline{X}_{p1} - \overline{X}_{p2}) \qquad (7\text{-}44)$$

where K is a constant of proportionality:

$$K = \frac{\sum_{p=1}^{k} \sum_{q=1}^{k} \lambda_p \lambda_q C_{pq}}{\sum_{p=1}^{k} \lambda_p (\overline{X}_{p1} - \overline{X}_{p2})}$$

Since the discriminant function has no absolute value, K can be given any convenient value. Here, K is given the value 1, such that:

$$\lambda_q = \sum_{p=1}^{k} C_{pq}^{-1} (\bar{X}_{p1} - \bar{X}_{p2}) \qquad (7\text{-}45)$$

and the discriminant function takes the form:

$$Z_{ij} = \sum_{q=1}^{k} \sum_{p=1}^{k} C_{pq}^{-1} (X_{p1} - \bar{X}_{p2}) \bar{X}_{qij} \qquad (7\text{-}46)$$

Since the population means and covariances are unknown, they must be estimated from the sample, giving:

$$\lambda_q = \sum_{q=1}^{k} C_{pq}^{-1} (\bar{X}_{p1} - \bar{X}_{p2}) \qquad (7\text{-}47)$$

and

$$Z_{ij} = \sum_{q=1}^{k} \sum_{p=1}^{k} C_{pq}^{-1} (\bar{X}_{p1} - \bar{X}_{p2}) X_{qij} \qquad (7\text{-}48)$$

A Probabilistic Extension

In mode-choice work the aim is not so much to classify individuals but more to predict the probability of a given individual's action so that the sum of such probabilities may be used as an estimate of the proportion of the population taking a given action. With this in mind, discriminant analysis was extended so that the discriminant score, Z, could be translated into a probability statement. Consider the frequency distributions of Z for both populations (Figure 7-3). The best estimate of the probability of an observation with a Z score of Z^* being from population 2 is:

$$p = \frac{N_2}{N_1 + N_2} \qquad (7\text{-}49)$$

Representing the frequency distributions as $F_1 (Z^*)$ and $F_2 (Z^*)$, the probability of being classified from population 2 at $Z^*, P (2 \mid Z^*)$ is:

$$P(2 \mid Z^*) = \frac{F_2(Z^*)}{F_1(Z^*) + F_2(Z^*)} = \frac{1}{1 + F_1(Z^*)/F_2(Z^*)} \qquad (7\text{-}50)$$

Now let $(Z_1 - Z_2) = 2d$, and let the origin be midway between the two means, i.e., $Z = z + t$, where t causes this shift. Recall that K in (7-42) was set equal to

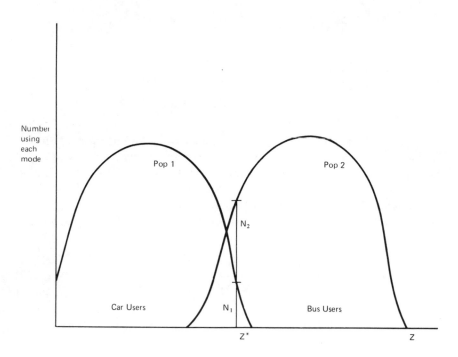

Figure 7-3. Frequency Distribution of Z-Scores

unity; it follows that σ^2 (the variance of each distribution) is equal to $2d$. Thus:

$$F_1(Z) = \frac{1}{\sqrt{2\pi\sigma^2}}\ e^{-1/2\left(\frac{(Z+d)^2}{\sigma^2}\right)};$$

$$F_2(Z) = \frac{1}{\sqrt{2\pi\sigma^2}}\ e^{-1/2\left(\frac{(Z-d)^2}{\sigma^2}\right)}; \tag{7-51}$$

$$P(2|Z) = \frac{1}{1 + e^{(-1/2((Z-d)^2-(Z+d)^2)/\sigma^2)}} \tag{7-52}$$

$$= \frac{1}{1 + e^{-2dZ/\sigma^2}} \tag{7-53}$$

Using $\sigma^2 = 2d$,

$$P(2|X) = \frac{1}{1 + e^{-Z}} \tag{7-54}$$

$$= \frac{1}{1 + e^{-(z+t)}} \qquad (7\text{-}55)$$

$$P(2|Z) = \frac{e^{(z+t)}}{1 + e^{(z+t)}} \qquad (7\text{-}56)$$

Note the similarity between this formulation and the logistic function. In this formulation it is assumed that nothing is known about the relative densities of the two populations, with the result that $Z = 0$ leads to $P(2 \mid Z = 0) = 0.5$. But the sample itself provides information on the relative proportion of population 1 and population 2 in the sample population. Given that N_1 and N_2 are the numbers in the two populations respectively, then the ordinate of $F_2(Z)$ will be increased by N_2/N_1:

$$F_2(Z) = \frac{N_2}{N_1} \frac{1}{\sqrt{2\pi\sigma^2}} e^{-1/2(z+d)^2/(\sigma)^2} \qquad (7\text{-}57)$$

$F_1(Z)$ remaining unchanged.
Therefore:

$$P(2|Z) = \frac{1}{1 + \dfrac{N_2}{N_1} e^{-2dZ/\sigma^2}} \qquad (7\text{-}58)$$

$$= \frac{\dfrac{N_1}{N_2} e^{(z+t)}}{1 + \dfrac{N_2}{N_1} e^{(z+t)}} \qquad (7\text{-}59)$$

Tests of Significance

The significance of the λ_p can be tested by using a standard t-test, and the significance of the function as a whole can be tested by means of an F-test. These procedures are sufficiently well known that it is unnecessary to treat them here.

A more unusual test of significance is the Mahalanobis D^2 statistic

$$D_2 = \sum_{p=1}^{k} \sum_{q=1}^{k} C_{pq}^{-1} (\overline{X}_{p1} - \overline{X}_{p2})(\overline{X}_{q1} - \overline{X}_{p1}) \qquad (7\text{-}60)$$

It is interpreted as the distance between the groups on a dimension that has unit standard deviation between the groups. No significant level of D^2 has been found in the literature search; it is normally assumed that the larger it is the better. However, the absence of a critical value for a given number of degrees of freedom leads to its rejection as unsatisfactory.

Evaluation of the Methods of Analysis

An effective evaluation of the four methods of analysis described above requires certain criteria. Aldana and Hoxie suggest that

four things seem important. First are the assumptions made by each type of analysis. Second are the quantity and quality of the data required by the techniques. Third is the computation time required. Finally, but most important, is the efficiency of the technique as measured by its "classificatory power" and the precision of the estimates rendered.[6]

For the purposes of selecting from the four methods the one most suitable for use in the analysis of transportation mode-choice models, these factors will be taken into account.

Assumptions and Properties

It has been shown above that the use of least-squares regression methods to estimate the linear probability function produces estimators that are, although unbiased, not efficient. Also, the facts that the errors are not normally distributed and that the presence of heteroskedasticity biases the standard errors of the coefficient estimates lead to problems in carrying out tests of significance. An appeal to asymptotic properties is required to deal with these problems, and even such an appeal cannot lead to efficient estimators. It is more important to note that the use of the linear probability function allows predictions of the dependent variable to run outside the unit interval, thus destroying the probability interpretation.

Probit analysis rests on the assumption that the threshold (or critical) values of the linear combination of independent variables are normally distributed. While it is impossible to test this assumption (since the threshold values are unobservable), two positions may be upheld. One asserts that the assumption of normality is unreasonable in the context of economic studies; the other argues that the assumption is reasonable if the sample size is adequate, and if the normality assumption is interpreted as reflecting the omission of some factors from the model.

Logit analysis uses a Bernouilli approach with the dependent variable made up of an unobservable probability and an unobservable error term. As in the case of probit analysis, this avoids the use of a stochastic error term and the necessity for assumptions about the distribution of that error term.

Discriminant analysis, by the use of a probabilistic extension of the basic technique, can be used to produce estimates of probabilities; but in this case it is no more than a special case of the logistic function in which the linear discriminant function is chosen as the most suitable linear combination. Moreover, the use of discriminant analysis leads to a more basic conceptual problem. Discriminant analysis was designed to solve classification problems, but in mode-choice models the aim is not to classify individuals as belonging to one group or another, but to estimate the probability that a traveler will choose one or other mode as a means to predicting the proportion of the sample which will choose a given mode. It is argued that proceeding from classification to the prediction of probabilities is less satisfying, and less efficient, than estimating the probabilities directly. It will later be argued that the results are also less good.

Given the above summary of the properties of each method, it remains to assess them in terms of their statistical properties. Discriminant analysis is less good than the others because it is both conceptually less suitable for the present task and a special, and more restrictive, case of the logistic function. It is not a reasonable procedure to indulge in additional effort to obtain a less general result.

It should be noted at this point that the remaining methods all require, to a greater or lesser degree, an appeal to the asymptotic properties of large samples. Thus, on this count no method is more suitable. It is argued that the inability of the linear probability function to restrict its predictions in the unit interval is a serious flaw, particularly as the aim of a mode-choice study is to predict the probabilities.

In terms of their statistical properties, it is difficult to distinguish between the probit and logit analyses. Both require large samples to ensure significant tests; both produce restricted estimates. In sum, it is suggested that logit is to be preferred, as it is an easier function to work with and the results are easier to interpret.

Data Requirements

Having assessed the statistical properties of the methods, it is now necessary to consider whether or not any of them has an advantage in terms of the amount of data required to estimate them. Since the least-squares regression, the probit, and the logit analyses appeal to asymptotic properties, they clearly require large samples. However, the use of discriminant analysis requires the assumption that the covariance matrices of each group be equal, and this is unlikely to be a valid assumption in a small sample. However, the importance of these restrictions depends on the definition of a "large sample." It is not possible to make firm statements about the size of sample required for the asymptotic properties to come into effect. However, it may be possible to make subjective statements about the minimum sample size required. Lisco has used probit analysis with 159 observations,[7] with results that have been confirmed by other analysts. It seems not unreasonable to postulate that all these methods may be estimated satisfactorily with one hundred observations, and it is hard to envisage any mode-choice study collecting data with less than this number of observations. (The analysis in Chapter 8 utilizes 1257 observations; clearly, this places no restrictions on the choice of method.) It may be concluded, therefore, that as far as data requirements are concerned no method is clearly more advantageous.

Prediction Capabilities

In order to test the performance of each method in terms of its ability to predict, an experiment was set up. It was decided to use each of the methods of analysis to estimate a model of mode choice, and to compare the predictions resulting from each of the methods. The model used is the model that proved to be the best in this study (Chapter 8), and took the following form:

$$P(T) = F \begin{Bmatrix} JU\,TRA \\ SUBCOS \\ WW\,TIM \\ TJT\,CA \end{Bmatrix}$$

$P(T)$ = probability of choosing train

$JU\,TRA$ = # of segments in train journey

$SUBCOS$ = cost of access to and egress from station

$WW\,TIM$ = walking/waiting time

$TJT\,CA$ = total journey time by car

The results of the estimations are presented in Table 7-1, supported by the means and standard deviations of the variables, presented in Table 7-2. Differences in the structures of the models estimated means that any direct comparison of the coefficients is a meaningless operation. Moreover, the four methods do not have common statistics; it is sufficient to note, however, that both the coefficients individually and the models taken as a whole are significant at the 0.99 significance level.

Since the models appear to produce estimates that are equally significant, a prediction test was proposed. The sample was divided randomly into two halves; the coefficients were estimated using one half of the data set, and these coefficients were used in conjunction with the other half of the data set to

Table 7-1
Means and Standard Deviations

Variable	Total		Coefs		Predic	
	Mean	S.D.	Mean	S.D.	Mean	S.D.
CHOICE (YY)	0.391	0.488	0.389	0.487	0.392	0.489
JU TRA	8.698	3.107	8.729	3.875	8.669	2.108
SUBCOS (£)	0.439	1.405	0.482	1.689	0.398	1.057
WW TIM (min)	34.128	21.252	34.668	34.668	21.141	21.362
TJT CA	102.417	50.946	104.924	57.343	99.976	43.741

Table 7-2
Coefficients Estimated on the COEFS Data Set

Variable	Mult. Reg.	Probit	Logit	Discrim
CONST.	0.5893	0.7590	1.3332	−4.1699
JU TRA	−0.0166	−0.1287	−0.2219	0.3656
SUBCOS	−0.0418	−0.2183	−0.3844	0.9231
WW TIM	−0.0052	−0.0019	−0.0205	0.1155
TJT CA	0.0014	0.0052	0.0090	−0.0306

produce estimates of the y vector. The appropriate test, then, is to compare the predictions of the y's which result from the different methods. Since the dependent variable is a conditional probability, the sum of the y's, in other words, the sum of the probabilities, will be the estimate of the number of people for whom the event will occur. In others words, Σy is equal to the number of people choosing the train. The results are presented in Table 7-3. It is interesting to note that the least-squares regression model produces the best result. It seems clear that the asymptotic properties of the O.L.S. regression model are being taken advantage of.

Discriminant analysis, however, performs less well than the other three models. An examination of the discriminant scores shows that the powers of this model to distinguish the car and train populations are low. Since the use of discriminant analysis relies on the classificatory approach, it is not surprising that a model that classifies poorly does not result in accurate predictions of the

Table 7-3
Results of the Prediction Test

Σy	$\Sigma \hat{y}$			
	Mult. Reg.	Probit	Logit	Discrim
250	249.00	244	246	292

probabilities. Since discriminant analysis may perform well in other cases, it is not possible to generalize completely and conclude that discriminant analysis is never appropriate. However, it is interesting to note that even when discriminant analysis does not perform well, probit and logit analyses produce estimates that result in very accurate predictions.

Lest it be thought that the test applied was biased against discriminant analysis, a classification test was also carried out. The probabilities for each traveler were predicted by each of the four methods, and the travelers were then assigned to either the train or the car on the basis of a $p = 0.5$ cut-off. The results are presented in Table 7-4.

The discriminant analysis misclassifies approximately twice as many observations as the other approaches. It is interesting to note that discriminant analysis has the lowest "net misclassification" figure; this means that the misclassifications offset each other and may account for the fact that discriminant analysis appears to perform well, since, on aggregate, the numbers predicted for each mode are close to the true values. However, it is obvious that this is accidental and conceals the fact that discriminant analysis has an extremely poor performance relative to the other methods. It should be noted that even when the discriminant scores were used to classify the observations, the discriminant analysis misclassified as follows:

Actual	Classified as	
	Car	Train
Car	241	138
Train	74	167

It is argued that these results support the point made above that even when discriminant analysis performs poorly, the probit and logit analyses produce predictions and classifications that are much more accurate.

Computational Requirements

While it is argued that the computational requirements should only be considered in selecting a method of analysis when all other things can be considered

Table 7-4
Misclassifications

	Mode 1	Mode 2	Total	Net
Probit	55	153	218	98
Logit	58	146	204	88
Discrim	218	183	401	35
Multreg	53	166	219	113

equal (which in this case they clearly cannot), the question is raised at this point for the sake of completeness in considering the criteria proposed by Aldana and Hoxie. They argue that probit or logit analysis takes three times as much execution time as discriminant analysis. The author's experience is somewhat different. For 621 observations and 4 independent variables, a program estimating probit and logit together and a discriminant program had the following characteristics:

	PROLO	DISCRIM
Central Processor Time (secs.)	23.416	4.586
Peripheral Processor Time (secs.)	27.109	10.714
Input/Output Time (secs.)	18.879	3.837
Total Time (secs.)	69.394	19.137
Cost	$4.09	$1.00

In this case both probit *and* logit take three and a half times as long as discriminant, and for each method the cost is only $2.045 compared with $1.00. It would appear that such a small difference is not sufficient to influence the choice of method.

Final Selection

In the light of the above discussions, one method must be selected for use in the analysis of the Edinburgh-Glasgow data. It is argued that the differences between the data and computational requirements do not provide any conclusive evidence to place any method at an advantage. Thus, the choice must be made on the basis of the statistical assumptions and properties and the performance characteristics of the methods. The fact that the O.L.S. regression method cannot restrict predictions between zero and one is held to be an objection serious enough to eliminate this method. The discriminant analysis is also rejected, for a number of reasons: first, it appears to the author to be an unsatisfactory approach from a conceptual standpoint; second, although its derivation is complex, the resulting probability function is a more restricted, special case of the logit function; and third, it performs badly on both prediction and classification tests.

Thus, the choice remains between probit and logit. Since they both require normality assumptions, have similar properties and virtually identical performances, the choice is a difficult one. Logit analysis is selected, however, as being somewhat easier to work with and to interpret.

It should be stressed, at this point, that the choice of logit analysis for this study reflects a complex process of trading off the good and bad points of each method. While each method has certain advantages, it is felt that, on balance, logit analysis is the most appropriate tool for use in studies of travel-mode choice.

8

The Analysis

In this study, considerable attention has been devoted to the problems both of the data collected in the Edinburgh-Glasgow Area Modal Split Study survey and the methods of analysis which could have been used to analyze this data. The special data-collecting problems and the unusual nature of the relationship to be estimated justified this rather lengthy procedure. The point has now been reached, however, when the analysis can be described. Following the conclusions of the previous chapter, logit analysis has been selected as the analytical tool, and it will be used to undertake three main investigations. The first will consider attempts to estimate a model using system characteristics alone. While such an investigation is clearly limited, the procedure is justified on the grounds that system characteristics are important to policy-makers. Moreover, a knowledge of the relative importance of the various system characteristics will render the addition of user characteristics a less cumbersome process, since some of the numerous variable combinations will be eliminated during the preliminary analysis. Thus, the second investigation will comprise the inclusion of the user characteristics. The third analysis will be devoted to an investigation of the effect of income in these models. Finally, the derivation of a value of time from the models estimated in this section will be examined.

Models of Systems Characteristics

In this section, eleven models will be presented. The variables included in these models represent combinations of variables and variable forms which reflect different hypotheses about the way in which the individual traveler assesses information of system characteristics. Inititally, no attempt will be made to compare these models in terms of their statistical efficiency, but they will all be closely examined to ensure that the models do not have undesirable implications. Specifically, it is important to consider the effects on the dependent variable of small changes in the independent variables. Should the inclusion of the variable be unjustified or the sign of the coefficient be incompatible with the underlying hypothesis, then changes in the independent variables will produce implausible effects on the dependent variable. Note that in all these models the dependent variable is the probability of choosing the train.

For ease of presentation, the variables are identified by their computer-acceptable names. A glossary of these names is presented in Table 8-1.

Table 8-1
Glossary of Computer Names

Name	Variable
CHOICE	Dichotomous dependent variable; 0 = car traveler, 1 = train traveler.
TIMDIF	Total journey time by train minus total journey time by car.
COSDIF	Total journey cost by train minus total journey cost by car.
JU DIF	Number of journey units by train minus number of journey units by car.
TJT TR	Total journey time by train.
TJC TR	Total journey cost by train.
TJC CA	Total journey time by car.
TJC CA	Total journey cost by car.
JU TRA	Journey units for the train journey.
AGE 1-5	Age dummy variables.
SEX 2	Sex dummy variable.
ADULTS	Number of adults in traveling party.
KIDS	Number of children in traveling party.
KIDRAT	Ratio of number of children to number of adults.
WW TIM	Walk/wait time.
SUBTIM	Subsidiary (i.e., nonline haul) transit time.
SUBCOS	Subsidiary (i.e., nonline haul) transit cost.
LHTIMD	Line-haul time time difference.
LHCOSD	Line-haul cost difference.
TD REL	Time different relative to overall journey time.
DC REL	Cost difference relative to overall journey cost.
WW REL	Walk/wait time relative to overall journey time.

The model chosen as the starting point of the analysis includes the time and cost difference variables that have been found to be suitable in models of commuter travel. Specification of the model purely in terms of times and costs is unnecessarily restrictive, and the proxy variables for inconvenience is therefore introduced at the outset of the analysis.

Model 1

Variable	Coefficient	t Value	Level of Significance
TIMDIF[1]	−.0039	−2.96	.01
COSDIF	−.0526	−1.26	I[a]
JU DIF	−.0835	−2.90	.01
CONST	.1779	1.12	I
Likelihood Ratio Test 33.34			.01

In this model all the coefficients have the hypothesized sign, in the sense that
[a]I indicates not significant at the .05 level.

the effects on the dependent variable produced by small changes in the independent variables are not contrary to those expected. Consider the time difference variable; the construction of this variable means that the following relationship exists:

$$P(T) = \alpha - \beta \, (\, TJT \, TR - TJT \, CA \,)$$

$P(T)$ = probability of choosing the train
$TJT \, TR$ = total journey time by train
$TJT \, CA$ = total journey time by car
α and β = constant parameters

Within this relationship, four changes can be considered, upward or downward changes in $TJT \, TR$ and $TJT \, CA$. If journey time by train should rise, then the probability of choosing the train would fall; if journey time by train should fall, then the probability of choosing the train would rise; if the journey time by car should rise, then the probability of choosing the train would also rise; and finally, if the journey time by car should fall, then the probability of choosing the train would fall. Since all these effects would be expected, it can be concluded that the sign of the coefficient estimated for the time difference variable is correct. Analogous reasoning confirms that the signs for the two other variables in this model are also correct.

Model 2

Variable	Coefficient	t Value	Level of Significance
TJT TR	−.0047	−3.62	.01
TJC TR	−.0239	− .48	I
TJT CA	.0079	4.671	.01
TJC CA	−.6497	1.97	.05
CONST.	−.6497	−4.01	.01
Likelihood Ratio Test 34.87			.01

Although it is commonly argued that the mode-choice decision is based on an examination of the difference in times and costs, it may also be hypothesized that it is the times and costs by each mode considered separately which form the decision base. Model 2 reflects this possibility, by including the variables in their original form.

Since the probability of choosing the train is affected positively by the car variables and negatively by the train variables, the signs of the coefficients are not unreasonable.

Model 3

Variable	Coefficient	*t* Value	Level of Significance
TJT TR	−.0025	−1.76	I
TJC TR	−.0258	−. 54	I
TJT CA	.0073	4.33	.01
TJC CA	.0942	1.66	I
JU DIF	−.1202	−3.89	.01
CONST	−.2489	−1.30	I
Likelihood Ratio Test	54.07		.01

Model 4

Variable	Coefficient	*t* Value	Level of Significance
TJT TR	.0013	.84	I
TJC TR	−.1275	−1.57	I
TJT CA	.0060	3.43	.01
TJC CA	.0787	1.28	I
JU DIF	−.0838	−2.75	.01
WW TIM	−.0234	−6.64	.01
CONST.	−.0066	.04	I
Likelihood Ratio Test	102.37		.01

Models 3 and 4 use the same basic data set as model 2 with the addition of two other variables, both of which can be regarded as proxy variables for the inconvenience of the journey by train. Both of these variables have the correct sign, since the probability of choosing the train is associated negatively with both the excess of journey units and the walking and waiting time associated with the train journey.

Model 5

Variable	Coefficient	*t* Value	Level of Significance
WW TIM	−.0246	−7.47	.01
SUBTIM	−.0034	−1.26	I
SUBCOS	−.3338	−3.71	.01
LHTIMD	.0016	1.17	I
LHCOSD	.0624	1.46	I
CONST.	.5328	3.96	.01
Likelihood Ratio Test	91.94		.01

Model 6

Variable	Coefficient	t Value	Level of Significance
WW TIM	−.0236	−7.07	.01
SUBTIM	−.0009	− .33	I
SUBCOS	−.3503	−3.90	.01
LHTIMD	.0014	.98	I
LHCOSD	.0577	1.38	I
JU DIF	−.0576	−2.06	.05
CONST.	.7370	4.44	.01
Likelihood Ratio Test	98.33		.01

It has been suggested that, particularly for a longer journey, the traveler may consider not simply the total times and costs, but rather the times and costs of the component parts of the journey. With this hypothesis in mind, models 5 and 6 include variables that reflect this, notably the walking and waiting time by train, the time and cost differences for both the line-haul and the access/egress sections of the journey. Model 6 adds the journey unit difference variable.

As expected, the probability of choosing the train varies negatively with the walking and waiting, the access and egress times and costs associated with the train journey. The positive signs on the line-haul time and cost differences reflect the fact that the variables were constructed in such a way that they normally take negative values. Thus, all the signs in these models are correct.

Model 7

Variable	Coefficient	t Value	Level of Significance
TD REL	−.6992	−6.14	.01
CD REL	−.1692	−2.82	.01
CONST.	−.0560	− .98	I
Likelihood Ratio Test	52.98		.01

Model 8

Variable	Coefficient	t Value	Level of Significance
TD REL	−.6522	−5.75	.01
CD REL	−.1503	−2.48	.05
JU DIF	−.0317	−2.69	.01
CONST.	.1055	1.27	I
Likelihood Ratio Test	60.79		.01

Model 9

Variable	Coefficient	t Value	Level of Significance
TD REL	−.4317	−3.59	.01
CD REL	−.1747	−2.86	.01
JU DIF	−.0212	−1.85	I
WW TIM	−.0107	−5.67	.01
CONST.	.3193	3.56	.01
Likelihood Ratio Test	93.41		.01

Model 10

Variable	Coefficient ·	t Value	Level of Significance
TD REL	−.3341	−2.72	.01
CD REL	−.2067	−3.34	.01
WW REL	−1.7725	−7.06	.01
JU DIF	−.0323	−2.73	.01
CONST.	.4685	4.79	.01
Likelihood Ratio Test	112.03		.01

As the object of this study is a medium-range, intercity trip, the possibility arises that the absolute value of the time and cost differences may be less important than the values of the differences relative to the total times and costs of the journey. It seems not unreasonable to argue that a time difference of five minutes is less important in a two-hour journey than in a ten-minute journey. The relative difference variables introduced in models 7, 8, 9, and 10 take this feature into account. The variables are constructed:

$$TD\ REL = \frac{(\ TJT\ TR\ -\ TJT\ CA\ )}{\dfrac{TJT\ TR\ +\ TJT\ CA}{2}}$$

(The denominator is set up as an average as a result of the difficulty of deciding which of the two alternatives was more suitable as a base.)

Thus, the relationship between the probability of choosing the train and the relative time difference is:

$$P(T) = \alpha - \beta \left(\frac{TJT\ TR - TJT\ CA}{TJT\ TR + TJT\ CA} \atop 2 \right)$$

Reasoning analogous to that employed in model 1 will confirm that the negative sign is correct. The signs of the other variables used in these models have been considered in earlier sections.

Model 11

Variable	Coefficient	t Value	Level of Significance
WW TIM	−.0203	−6.15	.01
TJT CA	.0096	6.36	.01
JU TRA	−.1893	−5.85	.01
SUBCOS	−.4552	−4.70	.01
CONST.	1.0237	3.71	.01
Likelihood Ratio Test	157.80		.01

Since the choice of behavioral hypothesis and, thus, of the variables in each model is based on a mixture of experience, casual empiricism, and introspection, it was decided that the infallibility of such a system was in doubt, and a check was instigated. All the variables of system characteristics were examined using stepwise regression analysis and stepwise discriminant analysis to seek out those variables with the strongest statistical relationship to the dependent variable. Both methods of analysis selected the same four variables, which give the above results when subjected to logit analysis. This reversal of the hypothesis-testing procedure requires particular care in the interpretation of the model, since it is easier to find hypotheses to fit statistical results than vice versa. Nevertheless, the interpretation of this model is most interesting.

The inclusion of the *TJT CA* variable with a positive sign is not unreasonable, since it implies simply that if the journey time by car increases, so will the probability of choosing the train. Each of the other three variables represents the negative effect of a disagreeable aspect of the train journey: the walking and waiting, the excess of journey units over the car journey, and the cost of travel to and from the station. Thus, this model may be interpreted as implying that for a longer journey, the traveler compares the absolute speed of the car with the sum of the inconveniences resulting from the journey by train.

Assessment of the Models

Having considered both the implications of the variable combinations represented by these models and the rationality of the signs of the estimated
coefficients, it is now necessary to assess the models in terms of their statistical
efficiency. The following criteria will be used: the first selection will be carried
out on the basis of the values of the likelihood ratios; further selection will
depend upon the level of significance attained by the individual variables in each
model.

At this point, a word of caution is appropriate: the nature of the likelihood
ratio test leads to a situation in which significance becomes easier to achieve as
the sample size increases. In this case, the sample size is so large that all the
models estimated are significant at the 0.01 significance level. Clearly, the
likelihood ratio test in this form is unreliable for indicating the relative
significances of a number of models. It is suggested, therefore, that the absolute
value of the likelihood ratio be used only as a guide to the performance of each
model. While it is acknowledged that the existence of stochastic elements
renders such a measure imperfect for distinguishing between values that are close
together, it provides a reasonable indication of the relative performances of the
various models.

In the light of this *caveat*, the models can be assessed. Ranking the models by
their likelihood ratios (Table 8-2) produces three distinct categories which stand

Table 8-2
Models Ranked by Likelihood Ratios

Model Number	Ratio
1	33.34
2	34.87
7	52.98
3	54.07
8	60.79
5	91.94
9	93.41
6	98.33
4	102.37
10	112.03
11	157.80

out even when the problems of ranking by likelihood ratio have been taken into account. On the basis of this ranking, Models 1, 2, 3, 7, and 8 will be rejected as being demonstrably less good than the other models. In contrast, Model 11 has a likelihood ratio so much greater than any of the others that, on this criterion, it must be selected as the best model. The remaining models are so close together in terms of their likelihood ratios that the secondary criterion must be applied to them. On this basis, Models 4, 5, 6, and 9 are rejected on the grounds that in each of them at least one variable was insignificant at the 0.05 level. Model 10 is acceptable since all the variables are significant at the 0.01 level.

Thus, out of the eleven models, Models 10 and 11 are selected as the best pair; nevertheless, Model 11 is regarded as the better of the two on the grounds that the likelihood ratio is so much greater.

Implications of the Assessment

The results of this assessment have two interesting implications. The first is that the models of time and cost differences successfully used in the analysis of the mode choices of commuters perform badly in models of the choices involved in longer journeys for noncommuting purposes. When a difference formulation is used successfully, it is a modified version that takes into account the fact that, for example, a five minute difference is less important on a long journey than on a short one. The second implication stems from the fact that the best model does not include a single variable in a difference formulation. The interpretation of Model 11 presented above would imply that what may be called the pseudo-system characteristics, such as walking/waiting time and inconvenience, are more important than the more obvious characteristics, such as times and costs.

Introduction of User Characteristics

In this section, the two best models of system characteristics will be used as a base, and to them will be added a number of user characteristic variables in order to investigate the extent to which the inclusion of such variables may improve the models. The variables to be added are age, sex, the number of adults in the party, the number of children in the party, and a composite variable of party size (# of children/# of adults). (The treatment of the income variable will be considered in the next section.) Since the effects of the user characteristics are the same when applied to both system models, the results of the augmented models will be presented in pairs.

Dummy Variables for Age

Variable	Coefficient	t Value	Level of Significance
TD REL	−.3509	−2.76	.01
CD REL	−.2529	−3.92	.01
WW REL	−1.9525	−7.50	.01
JU DIF	−.0362	−2.95	.01
AGE 1	−.8101	−6.35	.01
AGE 2	−1.1705	−8.34	.01
AGE 3	−1.2590	−8.69	.01
AGE 4	−1.1072	−7.33	.01
AGE 5	−1.1042	−5.12	.01
CONST.	1.4402	9.69	.01
Likelihood Ratio Test = 215.34			.01

Variable	Coefficient	t Value	Level of Significance
JU TRA	−.1130	−5.88	.01
SUBCOS	−.2572	−4.95	.01
WW TIM	−.0137	−6.97	.01
TJT CA	.0056	6.53	.01
AGE 1	−.8062	−6.23	.01
AGE 2	−1.1253	−7.94	.01
AGE 3	−1.2357	−8.45	.01
AGE 4	−1.0817	−7.08	.01
AGE 5	−1.0779	−4.89	.01
CONST.	1.5536	7.78	.01
Likelihood Ratio Test = 250.176			.01

These two models represent the inclusion of a series of dummy variables to take account of the age factor. The age dummy variables were constructed in the following way:

	Age 1	Age 2	Age 3	Age 4	Age 5
< 21	0	0	0	0	0
21-30	1	0	0	0	0
31-40	0	1	0	0	0
41-50	0	0	1	0	0
51-60	0	0	0	1	0
61 ⩽	0	0	0	0	1

Thus, the base is the "under 21 years old" group, and the signs on the coefficients indicate that if a subject falls into any of the "over 21 years" groups, then the probability that he will choose the train will be diminished. Given that, in Great Britain, few people under 21 years old would have access to a car, this result is not unreasonable. It is also interesting that the amount by which the probability is diminished is not constant for each age group; it increases up to the "41-50 years old" group and then decreases. This can be interpreted as meaning that up to that age group the subject is increasingly less likely to choose the train (as compared with the "under 21" group) but that this effect is diminished in the "51-60" and "61 and over" age groups.

Two factors may help to explain this result. The first is that the upper two age groups are more likely to include a number of people who, because of age and/or infirmity, are less likely to drive. However, such characteristics would also tend to make train travel difficult. The second factor is that the upper age groups may include many people who grew up before the automobile boom and thus never learned to drive or never acquired a car. Such an explanation is, of necessity, only partial, but these two factors may help to explain the diminished coefficients for the upper two age groups.

Dummy Variables for Sex

Variable	Coefficient	t Value	Level of Significance
TD REL	−.3640	−2.84	.01
CD REL	−.2625	−4.03	.01
WW REL	−1.9019	−7.26	.01
JU DIF	−.0399	−3.26	.01
AGE 1	−.7937	−6.15	.01
AGE 2	−1.0960	−7.72	.01
AGE 3	−1.1708	−8.00	.01
AGE 4	−1.0273	−6.71	.01
AGE 5	−.9959	−4.58	.01
SEX 2	.5329	5.73	.01
CONST.	1.2785	8.47	.01
Likelihood Ratio Test = 248.33			.01

Variable	Coefficient	t Value	Level of Significance
JU TRA	−.1209	−6.24	.01
SUBCOS	−.2551	−4.93	.01
WW TIM	−.0135	−6.82	.01
TJT CA	.0055	6.42	.01
AGE 1	−.7888	−6.03	.01
AGE 2	−1.0439	−7.29	.01
AGE 3	−1.1456	−7.76	.01
AGE 4	−.9980	−6.45	.01
AGE 5	−.9708	−4.37	.01
SEX 2	.5438	5.81	.01
CONST.	1.4447	7.20	.01
Likelihood Ratio Test = 284.13			.01

This pair of models includes a dummy variable to take account of the sex factor, the variable taking the value "0" for male subjects and "1" for female subjects. The results indicate that the probability of choosing the train is increased if the subject is female. The facts that, in Great Britain, fewer women drive and that it is safe for a woman to travel by train would seem to explain this result.

Variables of Party Size

Variable	Coefficient	t Value	Level of Significance
TD REL	−.3121	−2.40	.01
CD REL	−.2433	−3.69	.01
WW REL	−1.8070	−6.82	.01
JU DIF	−.0355	−2.85	.01
AGE 1	−.7989	−6.12	.01
AGE 2	−1.1807	−8.15	.01
AGE 3	−1.2196	−8.19	.01
AGE 4	−1.0619	−6.81	.01
AGE 5	−1.0116	−4.61	.01
SEX 2	.5916	6.28	.01
ADULTS	−.7184	−6.28	.01
CONST.	2.5666	9.90	.01
Likelihood Ratio Test = 288.62			.01

Variable	Coefficient	t Value	Level of Significance
JU TRA	−.1073	−5.44	.01
SUBCOS	−.2766	−5.04	.01
WW TIM	−.0129	−6.40	.01
TJT CA	.0053	6.11	.01
AGE 1	-.7969	−6.01	.01
AGE 2	−1.1325	−7.73	.01
AGE 3	−1.1974	−7.95	.01
AGE 4	−1.0412	−6.59	.01
AGE 5	−1.0009	−4.46	.01
SEX 2	.6043	6.36	.01
ADULTS	−.7412	−6.38	.01
CONST.	2.7303	9.42	.01
Likelihood Ratio Test = 325.85			.01

The addition of the variable representing the numbers of adults in the traveling party poses interpretational problems. The variable would seem to imply that as the number of adults in the party increases, the probability of choosing the train will diminish, but it is by no means intuitively obvious that this is a reasonable result. Moreover, the addition of a further variable representing the number of children in the party merely complicates the issue:

Variable	Coefficient	t Value	Level of Significance
TD REL	−.3108	−2.39	.01
CD REL	−.2436	−3.69	.01
WW REL	−1.8099	−6.83	.01
JU DIF	−.0349	−2.81	.01
AGE 1	−.7991	−6.12	.01
AGE 2	−1.1775	−8.13	.01
AGE 3	−1.2207	−8.20	.01
AGE 4	−1.0666	−6.84	.01
AGE 5	−1.0141	−4.62	.01
SEX 2	.5903	6.27	.01
ADULTS	−.7578	−6.26	.01
KIDS	.2375	1.01	I
CONST.	2.1694	4.61	.01
Likelihood Ratio Test = 289.64			.01

Variable	Coefficient	t Value	Level of Significance
JU TRA	−.1068	−5.41	.01
SUBCOS	−.2814	−5.06	.01
WW TIM	−.0128	−6.38	.01
TJT CA	.0054	6.16	.01
AGE 1	−.7967	−6.01	.01
AGE 2	−1.1292	−7.71	.01
AGE 3	−1.1976	−7.96	.01
AGE 4	−1.0458	−6.61	.01
AGE 5	−1.0025	−4.46	.01
SEX 2	.6029	6.34	.01
ADULTS	−.7866	−6.37	.01
KIDS	.2661	1.11	I
CONST.	2.2799	4.49	.01
Likelihood Ratio Test = 327.10			.01

Not only is the coefficient of the "number of children" variable statistically insignificant, but the sign is difficult to interpret, since it implies that the probability of choosing the train will increase if the number of children in the party increases. It is difficult to see why this should be so; indeed, it would be argued that the more children there are, the more likely it would be that the car would be chosen (subject, of course, to capacity constraints).

In order to test the possibility that the children and adult variables acted together, a composite variable was constructed that comprised the ratio of children to adults in the traveling party: $KIDRAT$ # of children in party/# of adults in party. Although the coefficient was statistically significant, its inclusion resulted in a reduction in the likelihood ratio, and its interpretation that the more children per adult the more likely it is that the train will be chosen is, at best, difficult to understand.

It is argued, therefore, that the variables of party size do not add anything to the models in the sense that either their coefficients are statistically insignificant or the variables are impossible to interpret.

Assessment of the Augmented Models

It is intended to assess the augmented models using the criteria previously used. The models involving party size variables are all rejected on the grounds that they do not improve the basic models. Thus, the problem of selecting which models to use for the investigation of the income effect resolves itself to one of deciding whether to add the age and sex variables. Since the series of age dummy variables are all highly significant statistically and, moreover, the addition of the five dummy variables increases the likelihood ratio substantially, from 112.03 to 215.34 and from 157.80 to 250.18, it is argued that such an increase represents an improvement in the model and that the age variables should, therefore, be included.

The decision as to the inclusion of the sex variable is, in effect, a value judgment, since it is not possible to state categorically that the increase in the likelihood ratio resulting from its addition is or is not significant. It has been decided, therefore, to include the variable.

The Effect of Income

For reasons set out in Chapter 4, it was decided not to include the income variable in the model, either as a dummy variable in the equation or as a multiplicative term attached to another variable, but to substratify the sample into income groups and to estimate the models separately for each income group. It can then be discovered whether any differences exist in the coefficients. Following the results presented in previous sections of this chapter, the models to be estimated will be:

System 1	System 2
TD REL	JU TRA
CD REL	SUBCOS
WW REL	WW TIM
JU DIF	TJT CA
AGE 1	AGE 1
AGE 2	AGE 2
AGE 3	AGE 3
AGE 4	AGE 4
AGE 5	AGE 5
SEX 2	SEX 2

The data set has been divided into five income groups; the details of each group are presented in Table 8-3. Each of the two model systems is estimated using each of the income groups, and the results are set out in Tables 8-4 and 8-5. It should be remembered that when these model systems were estimated using the complete data set, all the coefficients were significant at the 0.01 level; in these tables an "#" indicates that the coefficient is significant at only the 0.05 level and an "*" indicates that the coefficient is insignificant at this level. (Coefficients without these signs are significant at the 0.01 level.)

Behavior of Coefficient Estimates
Across Income Groups

The new estimates will be considered less in terms of their individual significance and more in terms of the implications of changes in significance across income groups. This being the case, the variables in each model system will be considered in order.

Table 8-3
Stratification by Income Group

Group	Limits (£)	Car Travellers		Train Travellers	Total
		#	%		
Y1	⩾ 1,000 p.a.	189	54.1	160	349
Y2	1,001-1,500 p.a.	118	37.2	199	317
Y3	1,501-2,000 p.a.	70	31.1	155	225
Y4	2,001-2,500 p.a.	30	22.9	101	131
Y5	2,501 ⩽	84	35.7	151	235
Total		491	39.1	766	1257

Table 8-4
System 1—All Income Groups

Variable	Y1	Y2	Y3	Y4	Y5
TD REL	−0.0914*	−0.0795*	−0.1736*		−1.3313
CD REL	−0.3611#	−0.1984*	−0.3047*	Failed to	−0.2078*
WW REL	−2.5356	−3.1118	−2.239	Reach A	−0.0490*
JU DIF	−0.0227*	−0.0207*	−0.0582	Maximum	−0.1336
AGE 1	−0.7520	−0.2336*	−3.7366	After 25	−1.3159#
AGE 2	−1.3658	−0.4271*	−3.8279	Iterations	−1.8346
AGE 3	−1.3936	−0.8277#	−3.4823*		−1.8225
AGE 4	−0.9075	−0.4807*	−4.3342*		−1.5850
AGE 5	−1.4071	−0.1634*	−3.8048*		−1.5227#
SEX 2	0.6199	0.5222	0.3819*		1.733*
Likelihood Ratio	84.68	62.15	32.90		63.06

Table 8-5
System 2—All Income Groups

Variable	Y1	Y2	Y3	Y4	Y5
JU TRA	−0.0501*	−0.0965	−0.1171#		−0.2258
SUBCOS	−0.3764	−0.2712	−0.3129*	Failed to	−0.2585
WW TIM	−0.0138	−0.0202	−0.0139	Reach A	−0.0075*
TJT CA	0.0064	0.0068	0.0075	Maximum	0.0025*
AGE 1	−0.7239	−0.3191*	−4.0283*	After 25	−1.0859*
AGE 2	−1.3578	−0.4787	−4.0687*	Iterations	−1.5059
AGE 3	−1.3088	−0.9601#	−3.7628*		−1.4621
AGE 4	−0.8020	−0.6335*	−4.5775*		−1.1497#
AGE 5	−1.3397	−0.1157*	−4.3086*		−1.0402*
SEX 2	0.6189	0.5745	0.4764*		−0.0463*
Likelihood Ratio	87.46	74.58	40.89		56.82

At this stage of the analysis, it is appropriate to make two points. First, the results cannot be considered complete, in the sense that no estimates are available for the fourth income group (Y4). The maximum likelihood procedure used to estimate this model failed to reach a maximum after twenty-five iterations. It is possible that this is due to either the small sample size or the small number of positive values (car travelers). Second, on only one occasion did a coefficient change sign (i.e., SEX 2 in System 1 − Y5); with this one exception, all the coefficients have the same sign as they did in the original estimations with the complete data set.

Model System 1

TD REL. It is interesting to note that only in the Y5 model is the relative time difference variable significant. This may be interpreted as indicating that only travelers in the upper income group place any importance on the travel time difference when making mode-choice decisions.

CD REL. Conversely, the relative cost difference variable is only significant in the Y1 model, (and then only at the 0.05 level), which can be given an analogous interpretation, i.e., that only travelers in the lowest income group consider the cost difference to be of importance; but even they do not attach much importance to it.

WW REL. The relative walking and waiting time variable is strongly significant in all models except that of the highest income group, which would imply that upper income group travelers do not consider walking and waiting time to be of importance. Such a contention may appear at first glance to be implausible, but a possible explanation is that such travelers, by the use of car, taxis, and limousines, minimize walking and waiting time in such a way that it is, in absolute terms, so small that it is unimportant. Thus, the effect demonstrated in the models may be the result of action taken because walking and waiting time is important, rather than a demonstration that it is unimportant.

JU DIF. It should be remembered that the journey unit difference variable is a proxy variable for inconvenience. Thus, the results imply that inconvenience is not considered to be an important factor for the lower income groups, and that it is only taken into account by the highest income group. Thus, it behaves much like the *TD REL* variable.

AGE. Even considering the dummy variables for age together, the combinations of different levels of insignificance do not reveal any trend across income groups. It is likely that the relationship between age and income is such that its effect is more complex than this analysis can reveal.

SEX. It is interesting that the importance of sex, both in terms of statistical significance and of the magnitudes of the coefficients, diminishes as income increases, resulting finally in a change of sign. Since it is more likely that a female in a higher income group would have access to a car, either as a passenger or as a driver, this result is not implausible.

Model System 2

JU TRA. The results indicate that the journey units by train variable (another proxy for the inconvenience of the train) is unimportant for the lower income group, and of increasing importance as income increases.

SUBCOS. The cost of access and egress variable maintains its significance (with the exception of model Y3). This is interpreted as meaning that the importance of this variable shows no consistent trend as income changes and is thus of equal importance to all income groups.

WW TIM. The original walking and waiting time variable also maintains significance except in the upper income group model (Y5). An explanation for this phenomenon has already been suggested in the remarks on *WW REL* in System 1.

TJT CA. The total journey time by car is similar to the *WW TIM* variable in its changes across income groups. It is more difficult, however, to explain the effect in this case, and, in fact, no plausible explanation has been found.

AGE/SEX. The remarks previously made with respect to AGE and SEX in Model System 1 apply equally here.

Modifications

As a result of the findings of the attempts to estimate separate income group models and in light of the discussion of the previous section, it has been decided that there is sufficient evidence to demonstrate that one model cannot be used for each income group. Thus, the models will be modified and the best combination of variables for each income group selected. The procedure will be to delete all variables whose coefficients proved to be insignificant and to include only tentatively those whose level of significance was reduced to the 0.05 level. The results will be presented for each income group.

Income Group 1

Variable	Coefficient	t Value	Level of Significance
CD REL	−0.3734	−2.59	.01
WW REL	−2.4768	−5.26	.01
AGE 1	−0.7556	−4.56	.01
AGE 2	−1.3901	−3.82	.01
AGE 3	−1.4035	−3.87	.01
AGE 4	−0.8988	−3.03	.01
AGE 5	−1.4182	−2.99	.01
CONST.	1.0902	5.38	.01
Likelihood Ratio Test = 84.24			.01

Variable	Coefficient	t Value	Level of Significance
SUBCOS	−0.3797	−2.97	.01
WW TIM	−0.0150	−4.63	.01
TJT CA	−0.0061	−3.68	.01
AGE 1	−0.7328	−4.41	.01
AGE 2	−1.4068	−3.79	.01
AGE 3	−1.3322	−3.61	.01
AGE 4	−0.7786	−2.64	.01
AGE 5	−1.3663	−2.88	.01
SEX 2	0.6169	4.06	.01
CONST.	0.4340	1.99	.01
Likelihood Ratio Test = 85.72			.01

As both of these models comprise variables that are all significant, other criteria must be used to select the better model. Moreover, since the likelihood ratio tests are almost identical, the value of the ratio cannot be used. This being the case, the first model is regarded as better, since it can achieve the same likelihood ratio with fewer variables. However, it must be acknowledged that there is little to choose between the models, and thus, should other considerations, such as ease of data collection, weight strongly, then either may be used.

Income Group 2

Variable	Coefficient	t Value	Level of Significance
WW REL	−3.1150	−6.12	.01
SEX 2	0.5143	2.92	.01
CONST.	0.5226	2.76	.01
Likelihood Ratio Test = 51.88			.01

Variable	Coefficient	t Value	Level of Significance
JU TRA	−0.0981	−2.66	.01
SUBCOS	−0.0981	−2.38	.05
WW TIM	−0.2357	−4.69	.01
TJT CA	0.0067	3.78	.01
SEX 2	0.5536	3.09	.01
CONST.	0.4426	1.33	I
Likelihood Ratio Test = 64.48			.01

As in the previous models for Income Group 1, it is argued that the likelihood ratios are insufficiently different to discriminate between the two models, and thus other more subjective criteria must be utilized. It is tempting to use the "principle of parsimony" and select the first model on the grounds that it requires only two variables, but the implication that the second income group only considers relative walking and waiting time is somewhat implausible. The second model should be selected, since it comprises the variables found to be the best set of explanatory variables in previous tests.

Income Group 3

Variable	Coefficient	t Value	Level of Significance
WW REL	−1.9807	−3.14	.01
CONST.	0.4339	1.49	I
Likelihood Ratio Test = 14.67			

Variable	Coefficient	t Value	Level of Significance
CD REL	−0.3289	−2.09	.05
WW REL	−2.3200	3.55	.01
CONST.	0.3478	1.17	I
Likelihood Ratio Test = 21.40			

Variable	Coefficient	t Value	Level of Significance
WW TIM	−0.0149	−3.18	.01
TJT CA	0.0041	1.96	.05
CONST.	−0.4053	−1.49	I
Likelihood Ratio Test = 15.26			

Variable	Coefficient	t Value	Level of Significance
JU TRA	−0.1060	−2.08	.05
WW TIM	−0.0134	−2.58	.01
TJT CA	0.0069	2.59	.01
CONST.	0.1960	0.44	I
Likelihood Ratio Test = 28.44			

The results for Income Group 3 are somewhat confusing; the likelihood ratios are much smaller, and the levels of significance change from model to model. While it is not unusual to find a change of significance when other variables are added, this phenomenon has not been apparent in the other models estimated. It is argued, therefore, that the models for this income group appear to be rather weak, and thus the fourth model is selected since it maximizes the likelihood ratio and, by the inclusion of *JU TRA*, brings the level of the two other variables up to the 0.01 significance level.

Income Group 4

No results are available for this income group, since the estimation procedure failed to reach a maximum after twenty-five iterations.

Income Group 5

Variable	Coefficient	t Value	Level of Significance
JU DIF	−0.1469	−3.53	.01
CD REL	−1.3298	−4.08	.01
AGE 1	−1.2346	−2.10	.05
AGE 2	−1.6620	−3.03	.01
AGE 3	−1.6668	−3.08	.01
AGE 4	−1.4265	−2.62	.01
AGE 5	−1.3634	−2.19	.05
CONST.	2.2265	−2.19	.05
Likelihood Ratio Test = 60.34			.01

Variable	Coefficient	t Value	Level of Significance
JU TRA	−0.2202	−5.73	.01
SUBCOS	−0.2098	−2.34	.05
CONST.	1.5064	4.65	.01
Likelihood Ratio Test = 39.55			.01

Variable	Coefficient	t Value	Level of Significance
JU TRA	−0.2260	−5.14	.01
SUBCOS	−0.2575	−2.60	.01
AGE 1	−1.0733	−1.86	I
AGE 2	−1.4943	−2.75	.01
AGE 3	−1.4508	−2.73	.01
AGE 4	−1.1393	−2.14	.05
AGE 5	−1.0364	−1.70	I
CONST.	2.7823	4.22	.01
Likelihood Ratio Test = 56.81			.01

In this set of models, the age variables were left in the System 1 reduced variable set, as only two of them failed to achieve significance at the higher level. In the model estimated with System 1 variables, they maintain the 0.05 level of significance, whereas in the models using System 2 variables, two of the age dummy variables are insignificant. Since the inclusion of age raises the likelihood ratio, the first model is selected as the best one for this income group.

The results of the model modifications are presented in Table 8-6. On the basis of these results the evidence is sufficiently conclusive to show that the effect of the income variable in models of model choice justifies the construction of different models for each income group. The models estimated for this data set indicate clearly that the models for each income group may be quite different. The conclusion is clear that the inclusion of income as a dummy variable or as a multiplicative term with other variables is unreasonable, since it conceals the more complex effects of the income variable.

The Derivation of a Value of Time

In Chapter 2 the procedure for deriving a value of time from a model of mode choice was developed, based on the premise that the desired value represented the change in cost required to just compensate for a one-unit change in time, where "just compensate" is interpreted as meaning that the net effect of the changes is to leave the probability unchanged. The derivation in Chapter 2 was set out in terms of cost and time variables expressed as differences. As the course of the analysis revealed, the time and cost difference variables were not satisfactory, in the sense that models using them were not statistically significant when estimated with the intercity data set. Thus, it is necessary to derive a value of time from the relative difference variables. The derivation is as follows.

It will be remembered that relationship between the probability of choosing the train and the relative time and cost differences is:

Table 8-6
The Selected Model for Each Income Group

Y1	Y2	Y3	Y4	Y5
CD REL	JU TRA	JU TRA		JU DIF
WW REL	SUBCOS	WW TIM		TD REL
AGE 1	WW TIM	TJT CA		AGE 1
AGE 2	TJT CA			AGE 2
AGE 3	SEX 2			AGE 3
AGE 4				AGE 4
AGE 5				AGE 5

$$P(T) = \frac{e^{G(x)}}{1 + e^{G(x)}} \qquad (8\text{--}1)$$

where G(x)

$$G(x) = \alpha_0 + \alpha_1 \left(\frac{(T_t - T_c)}{\left(\frac{T_t + T_c}{2} \right)} \right) + \alpha_2 \left(\frac{(C_t - C_c)}{\left(\frac{C_t + C_c}{2} \right)} \right) \qquad (8\text{--}1)$$

T_t = time by train
T_c = time by car
C_t = cost by train
C_c = cost by car

Since the $G(x)$ function is linear, the nonlinearities in the relationship may be ignored. Thus, the value of time is defined as the change in the cost variable required to just compensate for a one-unit change in the time variable. If the net result of the changes is to leave $G(x)$ unchanged, then $P(T)$ will also be unchanged.

The function can be made more manageable by simplifying to:

$$G(x) = \alpha_0 + 2\alpha_1 \left(\frac{T_t - T_c}{T_t + T_c} \right) + 2\alpha_2 \left(\frac{C_t - C_c}{C_t + C_c} \right) \qquad (8\text{--}2)$$

Taking the total differential with respect to T_t gives:

$$dG(x) = 2\alpha_1 \left(\frac{(T_t + T_c)dT_t - (T_t - T_c)dT_t)}{(T_t + T_c)^2} \right) \qquad (8\text{--}3)$$

$$= 2\alpha_1 \left(\frac{T_t dT_t + T_c dT_t - T_t dT_t + T_c dT_t}{(T_t + T_c)^2} \right) \qquad (8\text{--}4)$$

$$= \frac{2\alpha_1 2 T_c dT_t}{(T_t + T_c)^2} \qquad (8\text{--}5)$$

Taking total differentials with respect to T_c, C_t, and C_c gives, respectively:

$$dG(x) = \frac{- 2\alpha_1 2 T_t dT_c}{(T_t + T_c)^2} \qquad (8\text{--}6)$$

$$dG(x) = \frac{2\alpha_2 2 C_c dC_t}{(C_t + C_c)^2} \qquad (8\text{-}7)$$

$$dG(x) = \frac{-2\alpha_2 2 C_t dC_c}{(C_t + C_c)^2} \qquad (8\text{-}8)$$

It will be remembered that the value of time is defined as the change in cost required to just compensate for a one-unit change in time. Thus, if the differentials with respect to, for example, T_t and C_t, are summed and set equal to zero, i.e.,

$$\left[\frac{2\alpha_1 2 T_c dT_t}{(T_t + T_c)^2}\right] + \left[\frac{2\alpha_2 2 C_c dC_t}{(C_t + C_c)^2}\right] = 0 \qquad (8\text{-}9)$$

then the expression for dC_t can be derived:

$$dC_t = -\left(\frac{\alpha_1}{\alpha_2}\right)\left(\frac{T_c}{C_c}\right)\left(\frac{C_t + C_c}{T_t + T_c}\right)^2 dT_t \qquad (8\text{-}10)$$

If dT_t is set equal to one (for a one-unit change in time), then dC_t can be evaluated at the mean values of the other variables.

It will be noted that this value of time represents the change in the cost of the train which is required to compensate for a one-unit change in the train time. Clearly, the change in the car cost required to compensate for a one-unit change in train time can be derived, as can the changes in both train and car costs required to compensate for a change in car time. These expressions are, respectively:

$$dC_c = \left(\frac{\alpha_1}{\alpha_2}\right)\left(\frac{T_c}{C_t}\right)\left(\frac{C_t + C_c}{T_t + T_c}\right)^2 dT_t \qquad (8\text{-}11)$$

$$dC_t = \left(\frac{\alpha_1}{\alpha_2}\right)\left(\frac{T_t}{C_c}\right)\left(\frac{C_t + C_c}{T_t + T_c}\right)^2 dT_c$$

and $$dC_c = -\left(\frac{\alpha_1}{\alpha_2}\right)\left(\frac{T_t}{C_t}\right)\left(\frac{C_t + C_c}{T_t + T_c}\right)^2 dT_c \qquad (8\text{-}12)$$

Thus, four values of time can be derived. It will be clear that the consumer will, at the margin, equate these values, so that, for example, if the train time changes, he will be indifferent between making the compensating cost adjustment in terms of train cost or car cost. Consider, then, the effect of a change in the train time. The expression for the value of time reduces to the ratio of the coefficients multiplied by an adjustment factor. The expressions are:

$$dC_t = \frac{\alpha_1}{\alpha_2} (0.00476) \qquad (8\text{--}13)$$

and

$$dC_c = \frac{\alpha_1}{\alpha_2} (0.00813) \qquad (8\text{--}14)$$

The two adjustment factors are very similar and lead to values of time of £ 0.3954 per hour (dC_t) and £ 0.4644 per hour (dC_c), respectively, evaluated using the coefficients from the best augmented model.

Similarly, the effects of a change in the car time can be shown to be:

$$dC_t = \frac{\alpha_1}{\alpha_2} (0.00693) \qquad (8\text{--}15)$$

and

$$dC_c = \frac{\alpha_1}{\alpha_2} (0.00813) \qquad (8\text{--}16)$$

yielding values of time of £ 0.5766 per hour and £ 0.6762 per hour respectively. Once again, the two adjustment factors are similar, indicating that the cost adjustment can be made either to the car cost or the train cost.

While the pairs of adjustment factors (and thus, values of time) are similar, there is a distinguishable difference between the cost adjustment required to compensate for a car time change and that required for a train time change, and it may be argued that this indicates that car time and train time are valued differently. It is, however, difficult to assess the magnitude of the difference owing to the existence of stochastic elements in the derivation. It is argued, therefore, that the difference is not large enough to justify the strong conclusion that car time and train time are valued differently; it is conceded, however, that these results could be construed as indicating such a possibility.

Given this possibility, the value of time for train time can be derived by taking the average of the two values. Thus, the value is £ 0.4309 per hour. Similarly, the value of car time is £ 0.6264 per hour. The overall value of time can be calculated as the average of the value of train time and the value of car time, and is £ 0.5286 per hour. This value of time is considered to be the value of time spent traveling for nonbusiness and non-journey-to-work purposes in the Forth-Clyde Corridor.

It was indicated earlier in this study that one objective was to derive values of time for each income group, in order to examine the behavior of the value of time across income groups. It is unfortunate that it is not possible to achieve this objective, since the time and cost variables did not appear together in any of the models estimated for the separate income groups.

Conclusions

A number of conclusions can be drawn from this analysis.

1. The models of simple time and cost difference variables found to be satisfactory in commuting studies are not satisfactory for analyzing the choices of intercity, social and recreational travelers.

2. If difference variables are required, then the relative difference formulation is preferable.

3. The same model cannot be used for each income group, since the effect of income is so complex that it modifies not individual variables, but the actual selection of variables.

4. A value of time can be derived for the data set as a whole.

5. Values of time cannot be derived for each income group, since the necessary variables do not appear in the models for any of the income groups.

9   Conclusions

Having completed the analysis, it is now appropriate to consider what conclusions may be drawn from this study. Since the various steps in the analysis have been considered in some detail in the preceding chapters, the aim of this chapter will be to summarize rather than reiterate the findings.

Models and Economic Theory

An important conclusion of this study concerns the attempts to derive the rationale for a mode-choice model from the theory of consumer demand. The fact that it has been shown that behavioral models of mode choice are not inconsistent with the theory of consumer demand is considered to be an important result. It may be inferred from either a utility maximization or a cost minimization approach that the first-order conditions may be met by trading off time against goods. Thus, a model based on a hypothesized trade-off between time and cost is shown to be consistent with the theory of the consumer and need no longer be justified solely in terms of a hypothesis based on casual empiricism.

Data Collection

Before this study was undertaken, doubts were expressed as to the feasibility of collecting journey data sufficiently detailed to allow the analysis to be carried out. That such an undertaking has been carried out successfully is ample testimony to its feasibility. It is now clear that travelers can be motivated to complete a lengthy and complicated questionnaire.

Methods of Analysis

The problem of selecting an appropriate method for analyzing a mode-choice situation has troubled analysts for some time. The statistical comparison of the four available methods demonstrated the advantages and disadvantages associated with the use of each method. The comparison of the alternative methods resulted in a complex evaluation of the good and bad points of each method. On balance, logit analysis was deemed to be the most appropriate tool, and it was used in the analysis of the E.G.A.M.S.S. data.

Model Content

Perhaps the most interesting findings of this study concerns the content of the models that were estimated successfully. The starting point of the time and cost difference variables that had been used in commuting models of mode choice were found to be unsatisfactory. Modifications to the form of the difference variables, which converted them to relative differences, achieved some degree of success, but the best models included no variables in a difference formulation. In contrast, they contained variables that represented the inconvenience of the access and egress sections of the journey by train. It is concluded that for journeys of this type, in other words, medium-range, intercity social and recreational trips, the traveler is concerned more with comfort and convenience than with mere times and costs. This conclusion clearly illustrates the dangers encountered when models of commuting behavior are extended to other situations. Further, it is clear that methods for utilizing additional variables representing comfort, convenience, safety, and so forth should be developed.

The Effect of Income

One of the aims of this study was to consider the effects of income on the mode-choice decision. Thus, the better models were estimated using subsamples resulting from a stratification by income. The results indicate quite clearly that income strongly influences mode-choice decision behavior. That certain variables proved to be insignificant in the submodels was not unexpected; what was interesting was the very complex manner in which the stratification by income modified the models, resulting in a completely different model for each income group. The strength of the effect of income and the manner in which it operates indicate that the question of the treatment of the income variable requires further study.

The Value of Time

In the light of the fact that one of the primary aims of this study was to produce a value of time for intercity, social-recreational trips and to obtain different values of time for each income group, the results on the value of time are somewhat disappointing. Nevertheless, the results are extremely interesting. A value of time could not be derived from the best model, since the requisite time and cost variables did not appear in that model; nor could a value of time be derived from any of the income stratified submodels, for the same reason. This result is itself valuable, since it indicates that the derivation of a value of time from an apparent time-cost trade-off situation may not be universally valid, since, in some cases, what appears to be a trade-off situation may not be perceived as such by the travelers.

Nonetheless, a value of time has been derived, albeit from a suboptimum model. The value of £ 0.5286 (ten shillings and seven old pence) per hour, or 67.5 percent of the average wage rate,[a] is considerably higher than the values found in commuting studies. Considering only British results (to avoid exchange rate problems) and comparing the values as a percentage of the wage rate (to minimize the problems of intertemporal comparisons), the following results emerge:

Study	VOT as % of wage rate
Beesley	31-37 (*But* 42-50 for higher income group)
Stopher	42
Quarmby	21-25
L.G.O.R.U.	53.5
E.G.A.M.S.S.	67.5

Clearly, the commuting values have quite a high variance, and it is acknowledged that the final value may be somewhat unreliable, since it is derived from a suboptimal model. Nevertheless, the social-recreational value is sufficiently different to indicate that the use of a value of time derived from a commuting model in a noncommuting situation is a dangerous procedure and may lead to serious errors in the evaluation of time-savings.

One of the objectives of this study was to show that "commuting values of time" may not be universally valid. There is now evidence to justify this conclusion.

General Conclusion

This study was one of a series of studies designed to investigate mode-choice modeling and the valuation of time in noncommuting contexts, since it was suspected that the differences in the study situations might lead to differences in the models themselves and in the values of time derived from them. On the basis of this study, it may be concluded that this suspicion was fully justified; different situations require different modeling efforts, and attempts to transfer results from one area to another are fraught with danger.

[a]The average wage rate is calculated on the basis of income group midpoints and an assumed 2000-hour year.

Notes

Notes

Chapter 1
Introduction

1. Foster and Beesley (21).
2. Coburn, et al. (12).
3. See Harrison and Quarmby (30), section 1.1, and Harrison and Taylor (31) for a more comprehensive discussion of the valuation of working time saved.
4. Beesley (6).
5. Moses and Williamson (50), p. 256.

Chapter 2
The Development of the Model

1. Little (43), p. 31.
2. Hicks (33), p. 26.
3. Majumdar (46), p. 84.

Chapter 3
Behavioral Models and Economic Theory

1. Some of the material in this chapter has appeared in "Behavioural Models and Economic Theory" (83).
2. Warner (82), p. 3.
3. Quarmby (58), p. 302.
4. Lisco (42), p. 1.
5. Ibid., p. 3.
6. Haney (29), p. 26.
7. Lange (40), p. 178.
8. Haney (29), p. 35.
9. Ibid., p. 38.
10. Ibid., p. 39.
11. Ibid., p. 42.
12. Ibid., p. 44.
13. Becker (5), p. 496.
14. Ibid., p. 501.
15. Evans' notation has been changed to facilitate the development of the following section.

16. Evans (17), p. 3.
17. Ibid., p. 15.
18. De Donnea (16).
19. Ibid., p. 359.
20. Ibid., p. 371.

Chapter 4
Variables and Variable Forms

1. The use of time and cost variables leads to problems involving measurement. For a discussion of the advantages of perceived over measured data, see Watson (84 and 85).
2. Mercadel (49).
3. Lave (41), p. 73.
4. E.g. Lisco (42) and Quarmby (58).
5. E.g. Lave (41) and De Donnea (16).

Chapter 5
Data Collection

1. It is acknowledged that surveying traffic in a single direction is not an optimal procedure, based as it is on the assumption that over a period of time the traffic patterns in each direction will balance. It would have been difficult, however, in terms of both administrative cooperation and on-the-ground feasibility, to mount an operation covering traffic in both directions.

Chapter 7
The Method of Analysis

1. Goldberger (26), p. 245.
2. For a proof of the consistency of the estimates, see Kmenta (39), section 8.1, p. 250 *et seq*.
3. The exposition in this section is based on Tobin (77).
4. For a full explanation, see Tobin (77), pp. 7-9.
5. See Kmenta (39), pp. 461-62.
6. Aldana and Hoxie (2), p. 1.
7. Lisco (42).

Chapter 9
Conclusions

1. Beesley (5); Stopher (64); Quarmby (55); L.G.O.R.U. (41).

Bibliography

Bibliography

1. Aigner, D.J.: *Basic Econometrics*, Prentice-Hall, Englewood Cliffs, N.J., 1971.

2. Aldana, E. and Hoxie, P.: "Comparison of Classification Techniques for Estimating the Value of Time," M.I.T., (unpublished).

3. Anderson, T.W.: *Introduction to Multivariate Statistical Analysis*, Wiley & Sons, New York, 1958.

4. Arrow, Kenneth J.: "Utilities, Attitude and Choice: a review note," *Econometrica*, vol. 26, 1958.

5. Becker, G.S.: "A Theory of the Allocation of Time," *Economic Journal*, vol. 75, 1965.

6. Beesley, Michael E.: "The Value of Time Spent Travelling—Some New Evidence," *Economica*, vol. 32, 1965.

7. Beelis, W.R.: "Costs of Traffic Inefficiencies," *Proceedings of the Institute of Traffic Engineers*, 1952.

8. Berkson, J.: "Application of the Logistic Function to Bio-Assay," *Journal of the American Statistical Association*, vol. 39, 1944.

9. Charlesworth, G. and Paisley, J.L.: "The Economic Assessment of Returns From Road Works," *Proceedings of the Institute of Civil Engineers*, 1959.

10. Cherniak, N.: *Effects of Travel Impedance Cost*, Highway Research Board, Special Report # 56, 1959.

11. Claffey, P.: *Characteristics of Passenger Car Travel on Toll Roads and Comparable Free Roads*, Highway Research Board Bulletin # 306, 1961.

12. Coburn, T.M., Beesley, M.E., and Reynolds, D.J.: *The London-Birmingham Motorway-Traffic and Economics*, Road Research Technical Paper # 46, Great Britain, H.M.S.O., 1960.

13. Dawson, R.F.F.: *An Analysis of the Cost of Road Improvement Schemes*, Road Research Technical Paper # 50, Great Britain, H.M.S.O., 1961.

14. Dawson, R.F.F.: *Evaluation of Time of Private Motorists by Studying their Behaviour; Report on a Pilot Experiment*, Road Research Laboratory, Research Note # 3474, 1963.

15. Dawson, R.F.F. and Wardrop, J.G.: *Passenger Mileage by Road in Greater London*, Road Research Technical Paper # 59, Great Britain, H.M.S.O., 1962.

16. De Donnea, F.X.: "Consumer behaviour, transport mode choice and value of time: some micro-economic models," *Regional and Urban Economics, Operational Methods*, vol. 1, no. 4, February 1972.

17. Evans, A.W.: *A General Theory of the Allocation of Time*, Glasgow University (unpublished), 1969.

18. Great Britain, Department of Scientific and Industrial Research, (Road Research Laboratory), *Research on Road Traffic*, H.M.S.O., London, 1965.

19. Finney, D.J.: *Probit Analysis*, Cambridge University Press, Cambridge, England, 1947.

20. Foster, C.D.: *The Transport Problem*, Blackie, London, 1963.

21. Foster, C.D. and Beesley, M.E.: "Estimating the Social Benefit of Constructing an Underground Railway in London," *Journal of the Royal Statistical Society*, vol. 126, 1963.

22. Friedlander, A.: *The Interstate Highway System*, North Holland Publishing Co., Amsterdam, 1965.

23. Gillhespy, Norman R.: "The Tay Road Bridge—A Case Study in Cost-Benefit Analysis," *Scottish Journal of Political Economy*, vol. 15, 1968.

24. Glanville, Sir W.H.: "Economic and Traffic Studies," *Proceedings of the Institute of Civil Engineers*, vol. 15, 1960.

25. Glassborow, D.W.: "The Road Research Laboratory's Investment Criteria Examined," *Bulletin of the Oxford University Institute of Statistics*, 1960.

26. Goldberger, A.S.: *Econometric Theory*, Wiley & Sons, New York, 1964.

27. Gronau, Reuben: *The Value of Time in Passenger Transportation: the Demand for Air Travel*, National Bureau of Economic Research, New York, 1970.

28. Gwilliams, B.: *Transport and Public Policy*, George Allen and Unwin, London, 1964.

29. Haney, D.: *The Value of Time for Passenger Cars: An Experimental Study of Commuters' Values*, vol. I, Stanford Research Institute, Menlo Park, California, 1967.

30. Harrison, A.J. and Quarmby, D.A.: *The Value of Time in Transport Planning: A Review*, European Conference of Ministers of Transport, Round Table, 1969.

31. Harrison, A.J. and Taylor, S.J.: *The Value of Working Time in the Appraisal of Transport Expenditures—A Review*, Department of the Environment, Time Research Note # 16, 1970.

32. Hennes, R.G.: "Freeways and Suburbs," *Traffic Quarterly*, 1956.

33. Hicks, John R.: *A Revision of Demand Theory*, Clarendon Press, Oxford, 1965.

34. Hill, D.M. and H.G. Van Cube: *Development of a Model for Forecasting Travel Mode Choice in Urban Areas*, Highway Research Record # 38, 1963.

35. Hope, K.: *Methods of Multivariate Statistics*, University of London Press, 1968.

36. Johnson, B.: "Travel Time and the Price of Leisure," *Western Economic Journal*, 1966.

37. Johnston, J.: *Econometric Methods*, McGraw-Hill, New York, 1963.

38. Kain, J.F.: "A Contribution to the Urban Transportation Debate: An Econometric Model of Urban Residential and Travel Behaviour," *Review of Economic Statistics*, vol. 46, 1964.

39. Kmenta, J.: *Elements of Econometrics*, Macmillan, New York, 1971.

40. Lange, O.: "Foundations of Welfare Economics," *Econometrica*, vol. 10, 1942.

41. Lave, C.A.: *Modal Split Models*, Unpublished Ph.D. Dissertation, Northwestern University, 1969.

42. Lisco, T.E.: *The Value of Commuters' Travel Time—A Study in Urban Transportation*, Unpublished Ph.D. Dissertation, University of Chicago, Department of Economics, 1967.

43. Little, I.M.D.: *A Critique of Welfare Economics*, Oxford University Press, (2nd ed.), 1958.

44. Local Government Operation Research Unit, *Planning for the Work Journey*, L.G.O.R.U., Reading, England, 1970.

45. McGillivray, R.G.: "Demand and Choice Models of Mode Split," *Journal of Transport Economics and Policy*, vol. 4, 1970.

46. Majumdar, Tapas: *The Measurement of Utility*, Macmillan, London, 1961.

47. Malinvaud, E.: *Statistical Methods of Econometrics*, Rand-McNally, Chicago, 1966.

48. Marshall, A.: *Principles of Economics*, Macmillan, London, 1959.

49. Mercadel, M.: *Choice of Mode of Transport—Psychological Motivation and the Econometric Approach*, Third Round Table of the European Conference of Ministers of Transport, Paris, 1968.

50. Mishan, E.J.: "Theories of Consumer Behaviour—A Cynical View," *Economica*, vol. 21, 1961.

51. Mohring, H.: "Land Values and the Measurement of Highway Benefits," *Journal of Political Economy*, vol. 69, 1961.

52. Morrison, D.F.: *Multivariate Statistical Methods*, McGraw-Hill, New York, 1967.

53. Moses, L.N. and Williamson, H.F.: "Value of Time, Choice of Mode, and the Subsidy Issue in Urban Transportation," *Journal of Political Economy*, vol. 71, 1963.

54. Neter, J. and Maynes, E.S.: "Correlation Coefficient with 0, 1 Dependent Variable," *Journal of the American Statistical Association*, vol. 65, No. 330, 1970.

55. Oort, C.J.: "Evaluation of Travelling Time," *Journal of Transport Economics and Policy*, 1969.

56. Phillips, C.J.: *Valuing Travel Time—Some Implications of Recent Theory*, Department of the Environment, Time Research Note #10, 1969.

57. Pigou, A.C.: *Economics of Welfare*, Macmillan, London, (4th ed.), 1962.

58. Quarmby, D.A.: "Choice of Travel Mode for the Journey to Work," *Journal of Transport Economics and Policy*, 1967.

59. Rao, C.R.: *Advanced Statistical Methods in Biometric Research*, Wiley & Sons, New York, 1952.

60. Rao, C.R.: *Linear Statistical Inference and its Applications*, Wiley & Sons, New York, 1965.

61. Samuelson, Paul A.: "A Note on the Pure Theory of Consumer Behaviour," *Economica*, vol. 5, 1938.

62. Samuelson, Paul A.: "A Note on the Pure Theory of Consumer Behaviour—An Addendum," *Economica*, vol. 5, 1938.

63. St. Clair, G.P. and Lieder, N.: "Evaluation of the Unit Cost of Time and of the Strain and Discomfort Cost of Non-Uniform Driving," *Highway Research Board*, Special Report #56, 1969.

64. Sharp, C.J.: "The Choice between Cars and Buses on Urban Roads," *Journal of Transport, Economics and Policy*, 1967.

65. Stewart, M.: *For Whom the Bridge Tolls*, New Saltire, 1962.

66. Stopher, P.R.: "Predicting Travel Mode Choice for the Work Journey," *Traffic Engineering and Control*, vol. 9, 1968.

67. Stopher, P.R.: "Factors Affecting Choice of Mode Transport," Unpublished Ph.D. Thesis, University College, London, 1967.

68. Stopher, P.R.: "A Probability Model of Travel Mode Choice for the Work Journey," *Highway Research Record* #283, 1969.

69. Stopher, P.R.: "Transportation Analysis Methods," Northwestern University (unpublished), 1970.

70. Theil, H.: *Principles of Econometrics*, Wiley & Sons, New York, 1971.

71. Theil, H.: "On the Estimation of Relationships Involving Qualitative Variables," *American Journal of Sociology*, vol. 76, 1970.

72. Theil, H.: "Conditional Logit Specifications for the Multivariate Analysis of Qualitative Data in the Multiple Response Case," *Report 6945*, Center for Mathematical Studies in Business and Economics, University of Chicago, 1965.

73. Theil, H.: "The Explanatory Power of Determining Factors in the Multivariate Analysis of Qualitative Data," *Report 6946*, Center for Mathematical Studies in Business and Economics, University of Chicago, 1969.

74. Theil, H.: "A Multinomial Extension of the Linear Logit Model," *International Economic Review*, vol. 10, 1969.

75. Thomas, T.C.: *The Value of Time for Passenger Cars: An Experimental Study of Commuters' Values*, Stanford Research Institute, Menlo Park, California, 1967.

76. Tintner, G.: *Econometrics*, Wiley & Sons, New York, 1965.

77. Tobin, J.: "The Application of Multivariate Probit Analysis to Econometric Survey Data," *Cowles Foundation Discussion Paper* #1, 1955.

78. Tobin, J.: "Estimation of Relationships for Limited Dependent Variables," *Econometrica*, vol. 26, 1958.

79. Tucker, H.: *Highway Economics*, 1942.

80. U.S. Dept. of Commerce: *Modal Split—Documentation of Nine Methods for Estimating Transit Usage*, Bureau of Public Roads/Office of Planning, 1966.

81. Vaswani, R.: *The Value of Automobile Transportation and Time in Highway Planning*, Proceedings of the Highway Research Board #37, 1958.

82. Warner, S.L.: *Stochastic Choice of Mode in Urban Travel: A Study in Binary Choice*, Northwestern University Press, Evanston, 1962.

83. Watson, P.L.: "Behavioural Models and Economic Theory," Department of the Environment, *Time Research Note* # 16, London, 1970.

84. Watson, P.L.: *The Choice of the Data Base for Modal Split Modelling and the Valuation of Time*, Proceedings of the "Urban Traffic Model Research" Symposium, Planning and Transportation Research and Computation Co., London, 1970.

85. Watson, P.L.: *Some Problems Associated with the Time and Cost Data Used in Travel Choice Modelling and the Valuation of Time*, Highway Research Record, 1971.

86. West, M.H.: "Economic Value of Time Savings in Traffic," *Proceedings of the Institute of Traffic Engineers*, 1946.

Index

Index

About the Author

Peter L. Watson, born in Forfar, Scotland, in 1944, was educated at the University of Edinburgh, where he read modern languages and economics. Since 1970, he has been an Assistant Professor of Economics and a Research Associate in the Transportation Center at Northwestern University. He has served as a consultant to the National Academy of Engineers, the Government of Malaysia, and the World Bank. His primary research interests are in the area of travel demand analysis, including econometric models of travel demand, the value of time, and urban goods movements, and he has written a number of articles on these subjects.